Workbook
Progress in Mathematics

SADLIER-OXFORD

Rose Anita McDonnell

Catherine D. LeTourneau

Anne Veronica Burrows

Rita C. Smith

Theresa J. Talbot

Alice R. Martin

Francis H. Murphy

M. Winifred Kelly

with

Dr. Elinor R. Ford

Series Consultants

Tim Mason
Math Specialist
Palm Beach County School District
West Palm Beach, FL

Margaret Mary Bell, S.H.C.J., Ph.D.
Director, Teacher Certification
Rosemont College
Rosemont, PA

Dennis W. Nelson, Ed.D.
Director of Basic Skills
Mesa Public Schools
Mesa, AZ

Sadlier-Oxford
A Division of William H. Sadlier, Inc.
New York, New York 10005-1002
www.sadlier-oxford.com

Table of Contents

| Textbook Chapter | Workbook Page | Textbook Chapter | Workbook Page |

Problem-Solving Strategy: Guess and Test

Name _____

Date _____

Linda made a blue and purple bracelet with 14 beads.
There were 4 more blue beads than purple beads.
How many beads of each color did she use?

Guess: Test:

4 purple, 8 blue ⟶ 4 + 8 = 12

5 purple, 9 blue ⟶ 5 + 9 = 14

Linda used 5 purple and 9 blue beads.

Solve. Do your work on a separate sheet of paper.

1. Charles and Pam worked a total of 16 hours on a model car. Pam worked two hours less than Charles. How long did each child work?

2. Carol and Brad spent $9 at the store. Brad spent $3 more than Carol. How much did they each spend?

3. Paul made a bracelet with 14 red and yellow beads. There are 2 more red beads than yellow beads. How many beads of each color did Paul use?

4. Naoko made 13 bookmarks. She made one more brown bookmark than black. How many did she make of each color?

5. Jeremiah and Sara made a tower with 15 black and red blocks. There were 3 fewer black blocks than red blocks. How many blocks did they use of each color?

6. Alison worked for 12 hours on a project. She worked 2 more hours on Sunday than on Saturday. How many hours did she work each day?

1

Problem-Solving Strategy: Hidden Information

Name_____

Date_____

> Beth bought 1 dozen eggs.
> She used 5 eggs to make breakfast for her family.
> How many eggs were left?
> Hidden information: There are 12 eggs in a dozen.
> $12 - 5 = 7$
> She has 7 eggs left.

Solve. Do your work on a separate sheet of paper
Use the Table of Measures on page 484 of your textbook.

1. Andrew wanted to make bread. He had 1 pint of milk. The recipe called for 2 cups of milk. Did he have enough milk?

2. Ms. Allen had 2 pans that each held a half dozen muffins. How many muffins could she bake at one time?

3. Danielle needed to buy a can of juice that cost 89¢. She had one dollar. Did she have enough money?

4. Peter paid 2 quarters for a muffin. Jane paid for a drink with 4 dimes. Who spent more money?

5. Juanita wanted to make a sign for the Bake Sale. She had a board that was 1 foot long. She had 6 inches of ribbon. How much more ribbon did she need to trim the length of the board?

6. Joshua needed 4 quarters to play a video game. His father gave him a one-dollar bill. Could he exchange the dollar for enough quarters to play the game?

7. Rashawn needed a piece of yarn that measured 8 inches. He cut a piece that was 1 foot long. How much yarn was extra?

8. Allen had a quarter. He found a dime. How much money did he have?

Problem-Solving Strategy: Find a Pattern

What are the missing numbers in the pattern?

2, 4, 3, 5, 4, 6, 5, 7, 6, ___, ___

Try adding, and then subtracting.

2, 4, 3, 5, 4, 6, 5, 7, 6, 8, 7
  +2 −1 +2 −1 +2 −1 +2 −1 +2 −1

The rule is add 2, subtract 1.
The missing terms are 8 and 7.

Write the missing numbers in each pattern.

Write the rule.

1. 3, 6, 9, 12, 15, 18, ____, ____, ____, _____

2. 4, 3, 5, 4, 6, 5, 7, 6, 8, ____, ____, ____, ____, _____

3. 4, 8, 12, 16, 20, 24, ____, ____, ____, ____, _____

4. 10, 15, 16, 21, 22, 27, 28, ____, ____, ____, ____, _____

5. 2, 6, 4, 8, 6, 10, 8, ____, ____, ____, ____, _____

6. 3, 6, 7, 10, 11, 14, 15, ____, ____, ____, ____, _____

7. 5, 10, 9, 14, 13, 18, 17, ____, ____, ____, ____, _____

8. 1, 5, 7, 11, 13, 17, 19, ____, ____, ____, ____, _____

9. 25, 20, 21, 16, 17, 12, 13, ____, ____, ____, ____, _____

10. 12, 14, 13, 15, 14, 16, 17, ____, ____, ____, ____, _____

11. 3, 7, 6, 10, 9, 13, 12, ____, ____, ____, ____, _____

12. Write your own number pattern.
Ask a friend to continue it.

_____

 3

Problem-Solving Strategy: Make A Table

Name _____

Date _____

Megan helps her parents at their restaurant. She earned $3 on Friday. On Saturday, she earned double that amount. On Sunday, she earned double the amount she made on Saturday. If the pattern continued, how much did she earn on Monday?

Megan earned $24 on Monday.

Day	Amount Earned
Fri.	$3
Sat.	$6 ($3 + $3)
Sun.	$12 ($6 + $6)
Mon.	$24 ($12 + $12)

Solve. Do your work on a separate sheet of paper.

1. At the Cozy Cafe, Megan sold 2 cups of tea by 7:00. By 8:00, she sold one more cup than at 7:00. By 9:00, she sold two more cups than at 8:00. By 10:00, she sold three more cups than at 9:00, and so on. How many cups had she sold by 12:00?

_____

2. The cook made 3 orders of pancakes on Monday. He made double that amount on Tuesday. On Wednesday, he made double the amount he made on Tuesday, and so on. How many orders did he make on Thursday?

_____

3. Megan made name cards for a special party. She put 6 cards on the first table, 6 cards on the next table, and so on. How many cards would she need for 4 tables?

_____

4. On Wednesday, every fourth customer who comes in before 9:00 wins a free meal. If 21 people come in before 9:00, how many people win a free meal?

_____

5. Megan's father shakes hands with every second customer who comes in to the restaurant. If 20 people come in to the restaurant, how many hands does he shake?

_____

6. If Megan's father shakes hands with every third person, how many hands does he shake?

_____

Hundreds

Name _____

Date _____

				hundreds	tens	ones
				1	3	6

Standard Form   **Expanded Form**   **Word Name**
     136   100 + 30 + 6   one hundred thirty-six

Write the number in standard form.

1. two hundred thirty-nine _____   **2.** six hundred forty _____

3. four hundred sixty-one _____   **4.** nine hundred five _____

5. seven hundred _____   **6.** one hundred seven _____

7. 400 + 4 _____   **8.** 900 + 20 + 3 _____   **9.** 800 + 80 + 1 _____

10. 500 + 80 _____   **11.** 100 + 90 + 9 _____   **12.** 600 + 50 + 2 _____

Write the number in expanded form.

13. 701 = ____ hundreds ____ tens ____ ones = _____ + _____ + _____

14. 383 = ____ hundreds ____ tens ____ ones = _____ + _____ + _____

15. 490 = ____ hundreds ____ tens ____ ones = _____ + _____ + _____

Write the place and the value of the underlined digit.

16. 36<u>5</u> _____   **17.** <u>7</u>53 _____

18. 5<u>0</u>2 _____   **19.** 4<u>9</u>8 _____

20. <u>6</u>31 _____   **21.** 84<u>0</u> _____

Comparing and Ordering Numbers

Name_____

Date _____

<table>
<tr><td>Compare: 607, 670</td><td>Order: 607, 670, 624</td></tr>
</table>

h	t	o
6	0	7
6	7	0

Hundreds digits are the same.
Compare the tens: **0 < 7**
So 607 < 670.

Order: 607, 670, 624
607   Hundreds digits are
670   the same.
624   Compare the tens.
Least to Greatest: 607, 624, 670
Greatest to Least: 670, 624, 607

Compare. Write < or >.

1. 16 _____ 19   **2.** 25 _____ 20   **3.** 190 _____ 160

4. 25 _____ 32   **5.** 10 _____ 28   **6.** 84 _____ 48

7. 18 _____ 81   **8.** 17 _____ 13   **9.** 705 _____ 806

10. 561 _____ 565   **11.** 876 _____ 678   **12.** 908 _____ 980

Write in order from least to greatest.

13. 20, 80, 40 _____ _____ _____

14. 92, 76, 79 _____ _____ _____

15. 327, 486, 418 _____ _____ _____

16. 563, 569, 560 _____ _____ _____

Write in order from greatest to least.

17. 30, 90, 20 _____ _____ _____

18. 66, 18, 68 _____ _____ _____

19. 295, 614, 641 _____ _____ _____

20. 821, 824, 800 _____ _____ _____

PROBLEM SOLVING

21. Pamela has 525 stickers. Pierre has 550.
Who has more? _____

Skip Counting Patterns

Name_____

Date _____

Skip count by 2.

1 2 3 4 5 6 7 8 9 10 11 12 13 14 15

Skip count by 3.

1 2 3 4 5 6 7 8 9 10 11 12 13 14 15

Skip count by 3. Write the numbers.

1. Start at 39. End at 54. _____

2. Start at 60. End at 45. _____

Skip count by 4. Write the numbers.

3. Start at 40. End at 60. _____

4. Start at 96. End at 76. _____

Write the missing numbers.

5. 24, 26, ____, 30, 32, ____

6. 9, 11, 13, ____, ____, ____

7. 10, 15, ____, 25, ____

8. 35, 45, 55, ____, ____, ____

9. 13, 16, 19, ____, 25, ____, ____

10. 95, 90, 85, ____, ____, ____

PROBLEM SOLVING

11. I am an odd number between 22 and 29.
You say me when you skip count by 3.
What number am I?   _____

       7

Thousands

thousands	hundreds	tens	ones		Expanded Form	Standard Form
4	9	0	8	→	4,000 + 900 + 0 + 8 =	4,908

Read 4,908 as: four thousand, nine hundred eight.

Write the number in standard form.

1. 6 thousands 1 hundred 4 tens 6 ones _____

2. 9 thousands 0 hundreds 2 tens 0 ones _____

3. one thousand, four hundred, fifty-eight _____

4. eight thousand, thirty _____ **5.** six thousand, five _____

6. 1,000 + 900 + 30 + 1 _____ **7.** 7,000 + 0 + 20 + 3 _____

8. 2,000 + 800 + 0 + 8 _____ **9.** 5,000 + 0 + 80 + 0 _____

10. 3,000 + 0 + 0 + 3 _____ **11.** 9,000 + 500 + 80 + 7 _____

Write the place and the value of the underlined digit.

12. 9,025 _____ **13.** 3,641 _____

14. 6,511 _____ **15.** 7,064 _____

16. 8,059 _____ **17.** 5,918 _____

18. 1,704 _____ **19.** 2,546 _____

20. 4,956 _____ **21.** 9,870 _____

Ten Thousands and Hundred Thousands

Name_____

Date_____

Expanded Form: 300,000 + 40,000 + 6,000 + 300 + 20 + 1

Standard Form: 346,321

hundred thousands	ten thousands	thousands	hundreds	tens	ones
3	4	6,	3	2	1

Word Name: three hundred forty-six thousand, three hundred twenty-one

Complete.

1. 80,000 = _____ thousands

2. 95,000 = _____ thousands

3. 10,000 = _____ thousands

4. 73,000 = _____ thousands

5. 655,000 = _____ thousands

6. 838,000 = _____ thousands

7. 331,000 = _____ thousands

8. 444,000 = _____ thousands

Write the number in standard form.

9. 3 ten thousands _____

10. 9 hundred thousands _____

11. 20,000 + 4000 + 700 + 10 + 6 _____

12. 400,000 + 30,000 + 3000 + 200 + 0 + 4 _____

13. seventy-one thousand, four hundred fifty-six _____

14. seventy thousand, eight hundred ninety _____

15. six hundred seventy-one thousand _____

16. forty-two thousand, five hundred _____

Write the value of the underlined digit.

17. 945,<u>3</u>04 _____

18. <u>3</u>86,297 _____

19. 856,7<u>3</u>1 _____

20. 7<u>3</u>6,088 _____

21. 129,75<u>3</u> _____

22. 643,<u>3</u>900 _____

9

Comparing and Ordering Larger Numbers

Name _____

Date _____

Compare: 2517 ? 2137

Compare thousands: 2000 = 2000
Compare hundreds: 500 > 100

So 2517 > 2137.

Order from least to greatest:
6378, 6972, 3001

3000 < 6000 → 3001 is least.
6972 > 6378 → 6972 is greatest.
From least to greatest: 3001, 6378, 6972

Compare. Write < or >.

1. 6321 _____ 2814   **2.** 4228 _____ 2488   **3.** 8330 _____ 8333

4. 5432 _____ 5342   **5.** 6123 _____ 4151   **6.** 7251 _____ 7521

7. 2811 _____ 3811   **8.** 9242 _____ 9341   **9.** 6201 _____ 5775

10. 9429 _____ 942   **11.** 3075 _____ 3072   **12.** 1631 _____ 1637

Write in order from to least to greatest.

13. 6661, 2228, 7777 _____   _____   _____

14. 8000, 5000, 9000 _____   _____   _____

15. 3802, 3560, 4923 _____   _____   _____

16. 2468, 2408, 2458 _____   _____   _____

17. 1801, 1803, 1800 _____   _____   _____

Write in order from greatest to least.

18. 8325, 9325, 8625 _____   _____   _____

19. 4996, 4986, 4998 _____   _____   _____

20. 5008, 5010, 5009 _____   _____   _____

21. 2134, 2143, 2431 _____   _____   _____

Rounding Numbers

Name_____

Date _____

Round to the nearest thousand.

2100 is between 2000 and 3000.
4500 is between 4000 and 5000.

2100 is *nearer* to 2000 than 3000. Round *down* to 2000.

4500 is *halfway* between 4000 and 5000. Round *up* to 5000.

Write the tens each is between. Then round to the nearest ten.

1. 43 _____

2. 75 _____

3. 97 _____

4. 31 _____

5. 66 _____

6. 372 _____

7. 506 _____

8. 868 _____

9. 455 _____

Write the hundreds each is between. Then round to the nearest hundred.

10. 452 _____

11. 824 _____

12. 119 _____

13. 798 _____

14. 555 _____

15. 673 _____

16. 274 _____

17. 321 _____

18. 418 _____

19. 638 _____

20. 497 _____

21. 853 _____

Write the thousands each is between. Then round to the nearest thousand.

22. 4521 _____

23. 9267 _____

24. 1865 _____

25. 8650 _____

26. 3264 _____

27. 7202 _____

28. 2580 _____

29. 5999 _____

30. 6435 _____

31. 1662 _____

32. 2314 _____

33. 3537 _____

Use with Lesson 1–9, text pages 48–49.

11

Money
Less Than $1.00

Name _____

Date _____

What is the total amount?

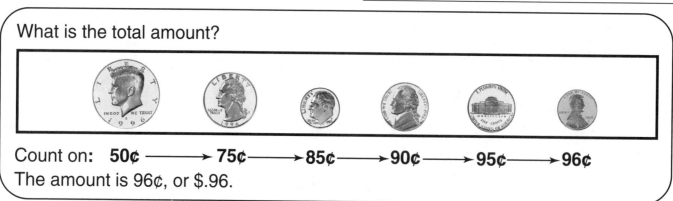

Count on: 50¢ ⟶ 75¢ ⟶ 85¢ ⟶ 90¢ ⟶ 95¢ ⟶ 96¢
The amount is 96¢, or $.96.

**Write the total amount in two ways. Use the cent sign.
Then use the dollar sign and the decimal point.**

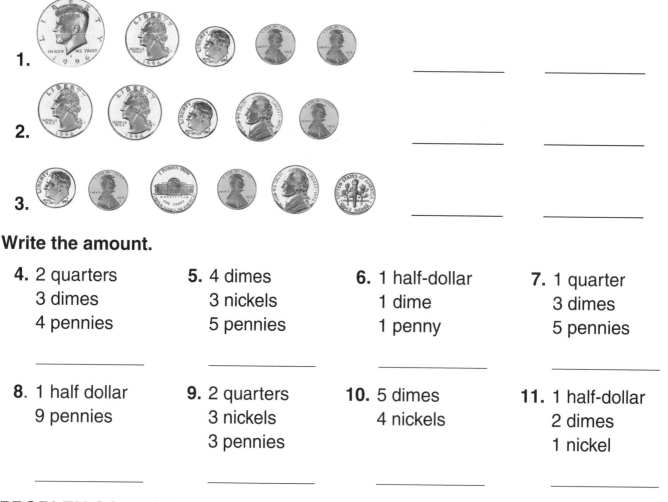

1. _____ _____

2. _____ _____

3. _____ _____

Write the amount.

4. 2 quarters
3 dimes
4 pennies

_____

5. 4 dimes
3 nickels
5 pennies

_____

6. 1 half-dollar
1 dime
1 penny

_____

7. 1 quarter
3 dimes
5 pennies

_____

8. 1 half dollar
9 pennies

_____

9. 2 quarters
3 nickels
3 pennies

_____

10. 5 dimes
4 nickels

_____

11. 1 half-dollar
2 dimes
1 nickel

_____

PROBLEM SOLVING

12. Maria went to the store. She spent 1 quarter, 2 dimes,
3 nickels, and 1 penny. How much did she spend? _____

Coins and Bills

Name_____

Date _____

Find the total amount.

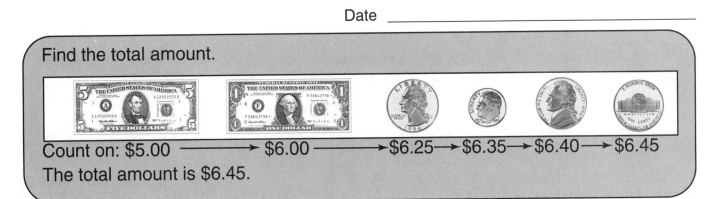

Count on: $5.00 ⟶ $6.00 ⟶ $6.25 → $6.35 → $6.40 → $6.45
The total amount is $6.45.

Write the amount. Use the dollar sign ($) and decimal point (.).

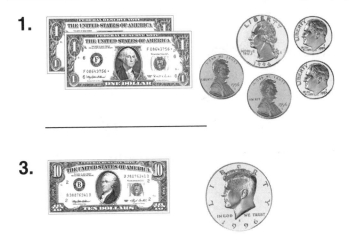

1. _____

2. _____

3. _____

4. _____

5. 3 one-dollar bills, 4 dimes, 3 nickels _____

6. 1 five-dollar bill, 1 quarter, 3 dimes _____

7. 1 ten-dollar bill, 5 dimes, 5 nickels _____

8. 4 ten-dollar bills, 1 half-dollar, 9 pennies _____

9. 2 five-dollar bills, 2 quarters _____

PROBLEM SOLVING

10. Lin wants to buy a book that costs $5.75.
She has 1 five-dollar bill and 4 quarters.
How much money does she have? _____

Making and Counting Change

Name_____

Date _____

Dana buys crayons for $3.65. She gives the cashier $5.00.
What is her change? How much is her change?

$3.65 $3.75 $4.00 $5.00

$\longrightarrow +10¢ \longrightarrow \longrightarrow +25¢ \longrightarrow \longrightarrow +\$1.00 \longrightarrow$

Arrange the money in order.
Count the change.
Her change is $1.35.

Ring the correct change.

1. Trena rode the Ferris wheel for 35¢. She gave the ticket collector $1.00.

   a.   b.

2. Fred spent $1.40 at the food booth. He gave the man 1 dollar, 2 quarters.

   a.   b.

PROBLEM SOLVING Write the change.

3. Jill gave the clerk $5.00 for her groceries. They cost $3.26. How much is her change?

   a. Use no quarters.

   _____

   b. Use no dimes.

   _____

4. Whitney buys an apple.
   She gives the cashier 2 quarters. _____

5. James buys yogurt.
   He gives the clerk $1.00. _____

6. Courtney buys a salad.
   She gives the cashier $3.00. _____

7. Paul buys a carton of milk.
   He gives the clerk 3 quarters. _____

Cafeteria Menu

Sandwich $2.25

Salad $2.49

Yogurt 75¢

Apple 29¢

Milk 55¢

Comparing and Rounding Money

Name _____

Date _____

Compare: $1.43 ? $1.49	Round $1.43 to the nearest dollar.
Compare dollars: $1.00 = $1.00	$1.43 is between $1.00 and $2.00.
Compare dimes: $.40 = $.40	$1.43 is *nearer* to $1.00.
Compare pennies: $.03 < $.09	Round $1.43 *down* to $1.00.
So $1.43 < $1.49.	

Compare. Write < or >.

1. $6.56 _____ $8.10   2. $3.98 _____ $2.98   3. $5.47 _____ $7.45

4. $4.77 _____ $4.67   5. $9.25 _____ $9.16   6. $1.35 _____ $1.25

7. $8.33 _____ $8.37   8. $6.74 _____ $6.54   9. $4.58 _____ $4.52

10. $3.77 _____ $3.79   11. $5.92 _____ $5.90   12. $2.63 _____ $2.64

13. $8.68 _____ $8.69   14. $7.29 _____ $7.39   15. $1.56 _____ $6.51

Round to the nearest dollar.

16. $6.51 _____   17. $1.73 _____   18. $4.39 _____   19. $8.25 _____

20. $3.34 _____   21. $7.58 _____   22. $2.95 _____   23. $5.08 _____

24. $9.16 _____   25. $6.64 _____   26. $1.39 _____   27. $4.52 _____

PROBLEM SOLVING

28. Ramon earned $8.75 raking leaves.

    Susie earned $8.35 raking leaves.

    a. Who earned more?　　　　　　　　　　 _____

    b. Round each amount to the nearest
       dollar.　　　　　　　　　　　　　　　　 _____

Problem-Solving Strategy: Draw A Picture

Name_____

Date _____

Allan is using 22 squares of material to make a small quilt.
Every 4th square is red. How many squares are red?
Draw 22 squares. Mark every 4th square. Count the number of squares you mark.

☐☐☐🅇☐☐☐🅇☐☐☐🅇☐☐☐🅇☐☐☐🅇☐☐

Five squares are red.

Solve. Do your work on a separate sheet of paper.

1. Jenny is making a bracelet with 20 beads. Every 5th bead is purple. How many purple beads will she need?

   _____

2. Carmen had 24 flowers. She put 3 flowers in each vase. How many vases did she use?

   _____

3. Don is planning a party for 6 friends. He will serve each friend 2 tuna sandwiches. How many sandwiches does he need to make for his friends?

   _____

4. Isabel's father gave her 5 dimes, 1 nickel, and 5 pennies. How much money did he give her?

   _____

5. Isaiah is making a pattern with the same number of red, white, blue, and green triangles. He uses 32 triangles. How many does he use of each color?

   _____

6. Yushiro made a tower with 25 red and black blocks. He started with a red block. Every other block was red. How many red blocks did he use?

   _____

7. Ben saved $2.41. Then he found 3 nickels and 9 pennies under his bed. How much money does he have now?

   _____

8. Jasmine helped her mother put 2 sheets of paper in each of 8 envelopes. How many sheets of paper did she use?

   _____

Adding: No Regrouping

Name _____

Date _____

Add: 6 + 4 + 5 + 1 = ?
Add down.

```
  6
  4      6 + 4 = 10
  5      10 + 5 = 15
+ 1      15 + 1 = 16
 16
```

Find the missing addend:
2 + ? = 5

2 + 3 = 5

Add: 32 + 24 = ?
First add the ones.
Then add the tens.

```
  32
+ 24
  56
```

Add.

1. 7	**2.** 2	**3.** 6	**4.** 4	**5.** 5	**6.** 1	**7.** 2
3	8	1	4	3	9	6
+ 5	+ 7	+ 9	6	2	5	8
			+ 3	+ 5	+ 3	+ 1

Write the missing addend. You may use a number line.

8. 7 + _____ = 13

9. _____ + 4 = 10

10. 4 + _____ = 12

11. 18 = 9 + _____

12. 14 = _____ + 6

13. 12 = 7 + _____

14. 6 + _____ = 11

15. _____ + 7 = 15

16. 9 = 5 + _____

Write the sum.

17. 7 2	**18.** 6 5	**19.** 8 2	**20.** 3 2	**21.** 1 2
+ 1 4	+ 2 0	+ 1 7	+ 4 4	+ 8 6

22. 6 1 2	**23.** 4 3 7	**24.** 1 3 9	**25.** 2 4 7	**26.** 7 6 5
+ 3 7 5	+ 2 6 1	+ 8 6 0	+ 3 5 1	+ 1 2 3

Use with Lessons 2–1, 2–2, and 2–3, text pages 68–73.

Estimating Sums

Name_____

Date _____

Round to the nearest:

Ten	Ten cents	Hundred	Dollar
49 ⟶ 50	\$.43 ⟶ \$.40	238 ⟶ 200	\$8.45 ⟶ \$8.00
+33 ⟶ +30	+ .26 ⟶ + .30	+159 ⟶ +200	+ 7.55 ⟶ + 8.00
about 80	about \$.70	about 400	about \$16.00

Estimate by rounding to the nearest ten or ten cents.

1. 5 6	2. 2 8	3. 3 2	4. 1 2 1
+1 3	+2 2	+6 4	+ 4 6

5. \$ .9 3	6. \$.5 6	7. \$.4 7	8. \$2.4 4
+ .2 2	+ .4 8	+ .6 3	+ .2 7

Estimate by rounding to the nearest hundred or dollar.

9. 5 6 1	10. 8 2 1	11. 1 7 7	12. 7 5 2
+2 1 3	+1 3 7	+3 4 6	+3 9 5

13. 7 2 7	14. 6 1 3	15. 2 7 1	16. 5 3 2
+1 6 2	+2 2 6	+6 2 7	+8 6 4

17. \$7.2 3	18. \$7.8 8	19. \$6.1 5	20. \$4.9 5
+ 8.8 4	+ 2.9 9	+ 2.3 6	+ 5.3 9

PROBLEM SOLVING Estimate each answer.

21. A carton of milk costs \$1.36 and a bunch of celery costs \$.98. **About** how much do the two items cost? _____

22. At the school play, the gym was filled with 232 parents and 57 students. **About** how many people came to the play? _____

Adding Money

Name_____

Date _____

First estimate. Then add.

12¢ → 10¢ 12¢	$6.64 → $7.00 $6.64
+35¢ → +40¢ +35¢	+ 1.22 → + 1.00 + 1.22
about 50¢ 47¢	about $8.00 $7.86

Estimate. Then add.

1. 3 5¢
  +6 2¢

2. 1 4¢
  +1 5¢

3. 2 5¢
  +6 1¢

4. 2 3¢
  + 4¢

5. 1 0¢
  + 8¢

6. 6 4¢
  +2 0¢

7. 4 3¢
  +1 0¢

8. 4 4¢
  + 5¢

9. 3 4¢
  +4 3¢

10. 8 1¢
  +1 8¢

11. $5.2 4
  + 1.0 2

12. $6.1 4
  + 2.8 3

13. $7.0 5
  + 2.4 2

14. $1.9 2
  + 7.0 7

15. $6.2 7
  + 3.6 1

16. $5.2 4
  + 3.1 3

17. $.7 2
  + .2 4

18. $9.2 8
  + .7 1

19. $.5 0
  + .3 7

20. $6.2 9
  + 2.3 0

PROBLEM SOLVING

21. Terry bought a stuffed monkey for $4.49 and a toy poodle for $3.50. How much did she pay for both?

_____

22. Carl called his parents from camp. The first call cost $2.66. The second cost $1.32. How much did he spend on both calls?

_____

Regrouping Ones

Name_____

Date _____

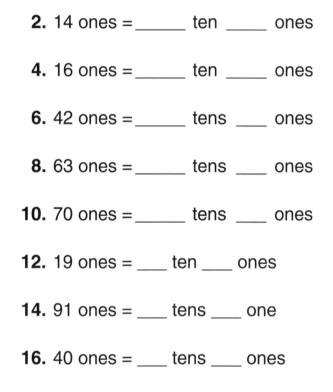

24 ones = ? tens ? ones

So 24 ones = 2 tens 4 ones.

Complete. You may use base ten blocks

1. 11 ones = _____ ten _____ one

2. 14 ones = _____ ten _____ ones

3. 18 ones = _____ ten _____ ones

4. 16 ones = _____ ten _____ ones

5. 37 ones = _____ tens _____ ones

6. 42 ones = _____ tens _____ ones

7. 81 ones = _____ tens _____ one

8. 63 ones = _____ tens _____ ones

9. 55 ones = _____ tens _____ ones

10. 70 ones = _____ tens _____ ones

11. 15 ones = _____ ten _____ ones

12. 19 ones = _____ ten _____ ones

13. 28 ones = _____ tens _____ ones

14. 91 ones = _____ tens _____ one

15. 32 ones = _____ tens _____ ones

16. 40 ones = _____ tens _____ ones

PROBLEM SOLVING

17. Lamar is playing a trading game with base ten blocks. He has 52 ones units. If he trades them for the fewest blocks, how many tens rods will he get? How many ones units? _____

Adding with Regrouping

Name _____

Date _____

Estimate.	Add.	tens	ones	tens	ones

Estimate.
26 → 30
+18 → +20
about 50

	tens	ones
	¹2	6
+	1	8
		4

	tens	ones
	¹2	6
+	1	8
	4	4

14 ones = 1 ten 4 ones

Estimate. Then find the sum.

1. 16
   +17

2. 29
   +41

3. 64
   +26

4. 72
   +19

5. 48
   +36

6. 54
   +29

7. 38
   +58

8. 49
   +36

9. 28
   +52

10. 24
    +68

11. 14
    +45

12. 22
    + 7

13. 38
    +18

14. 36
    +57

15. 49
    +21

16. 67
    + 8

17. 29
    + 5

18. 47
    +27

19. 19
    +21

20. 14
    +68

21. 17¢
    +18¢

22. 66¢
    +24¢

23. 25¢
    +58¢

24. 52¢
    +18¢

25. 46¢
    + 5¢

PROBLEM SOLVING

26. How many blue and green erasers are there in all? _____
27. Green and yellow? _____
28. Yellow and red? _____
29. Blue and yellow? _____
30. Green and red? _____

Erasers			
BLUE	GREEN	YELLOW	RED
24	17	18	19

Regrouping Tens

Name _____

Date _____

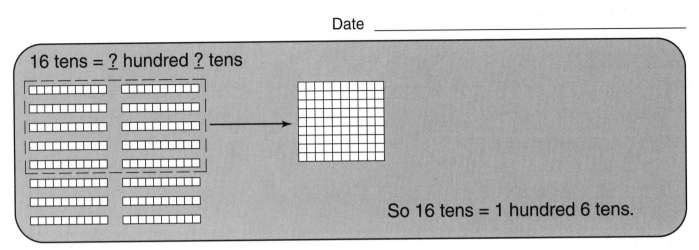

16 tens = <u>?</u> hundred <u>?</u> tens

So 16 tens = 1 hundred 6 tens.

Complete. You may use base ten blocks.

1. 14 tens =

_____ hundred _____ tens

2. 17 tens =

_____ hundred _____ tens

3. 25 tens =

_____ hundreds ___ tens

4. 64 tens =

_____ hundreds ___ tens

5. 32 tens =

_____ hundreds ___ tens

6. 41 tens =

_____ hundreds ___ tens

7. 50 tens =

_____ hundreds ___ tens

8. 10 tens =

_____ hundred _____ tens

9. 76 tens =

_____ hundreds ___ tens

10. 83 tens =

_____ hundreds ___ tens

11. 97 tens =

_____ hundreds _____ tens

12. 28 tens =

_____ hundreds _____ tens

13. 59 tens =

_____ hundreds _____ tens

14. 44 tens =

_____ hundreds _____ tens

15. 60 tens =

_____ hundreds _____ tens

16. 80 tens =

_____ hundreds _____ tens

Adding: Regrouping Tens

Name _____

Date _____

Add the ones.	Add the tens.	Regroup.

	h	t	o
		6	8
+		5	1
			9

	h	t	o
		6	8
+		5	1
	1	1	9

11 tens = 1 hundred 1 ten

Complete.

1. $\begin{array}{r} 9\,3 \\ +3\,4 \\ \hline 7 \end{array}$

2. $\begin{array}{r} 4\,5 \\ +8\,2 \\ \hline 1\quad7 \end{array}$

3. $\begin{array}{r} 8\,2 \\ +3\,7 \\ \hline 1 \end{array}$

4. $\begin{array}{r} 4\,3 \\ +6\,5 \\ \hline 1 \end{array}$

5. $\begin{array}{r} 2\,6 \\ +8\,1 \\ \hline 1 \end{array}$

Add.

6. $\begin{array}{r} 5\,6 \\ +7\,1 \\ \hline \end{array}$

7. $\begin{array}{r} 8\,4 \\ +4\,5 \\ \hline \end{array}$

8. $\begin{array}{r} 9\,2 \\ +2\,2 \\ \hline \end{array}$

9. $\begin{array}{r} 5\,5 \\ +8\,3 \\ \hline \end{array}$

10. $\begin{array}{r} 7\,3 \\ +3\,5 \\ \hline \end{array}$

11. $\begin{array}{r} \$.5\,2 \\ +\ .8\,6 \\ \hline \end{array}$

12. $\begin{array}{r} \$.7\,0 \\ +\ .5\,5 \\ \hline \end{array}$

13. $\begin{array}{r} \$.6\,3 \\ +\ .8\,2 \\ \hline \end{array}$

14. $\begin{array}{r} \$.7\,1 \\ +\ .4\,2 \\ \hline \end{array}$

15. $\begin{array}{r} \$.5\,4 \\ +\ .9\,5 \\ \hline \end{array}$

16. $\begin{array}{r} 4\,0 \\ +7\,9 \\ \hline \end{array}$

17. $\begin{array}{r} 2\,3 \\ +9\,5 \\ \hline \end{array}$

18. $\begin{array}{r} 7\,2 \\ +8\,2 \\ \hline \end{array}$

19. $\begin{array}{r} 4\,5 \\ +6\,4 \\ \hline \end{array}$

20. $\begin{array}{r} 5\,3 \\ +8\,5 \\ \hline \end{array}$

PROBLEM SOLVING

21. Charles collected 54 cans on Friday for the recycling drive. He collected 62 cans on Saturday. How many cans did he collect altogether?

_____

Adding: Regrouping Twice

Name _____

Date _____

Estimate.

$$92 \longrightarrow 90$$
$$+39 \longrightarrow +40$$
about 130

Add.

h	t	o
	1	
	9	2
+	3	9
		1

11 ones =
1 ten 1 one

h	t	o
	1	
	9	2
+	3	9
1	3	1

13 tens =
1 hundred 3 tens

Complete.

1. 99
   +34
    3

2. 57
   +75
   1 3

3. 63
   +39
   1 2

4. 88
   +52
   1

5. 45
   +65
   1 0

Estimate. Then add.

6. 78
   +24

7. 56
   +64

8. 47
   +58

9. 79
   +65

10. 58
    +47

11. 82
    +48

12. 64
    +38

13. 49
    +56

14. 72
    +68

15. 85
    +36

16. 45
    +59

17. 97
    +87

Find the sum.

18. 44
    +56

19. 29
    +92

20. 88
    +76

21. 64
    +76

22. 46
    +57

23. 49
    +52

24. $86 + 57 =$ _____

25. $55 + 67 =$ _____

PROBLEM SOLVING

26. Mr. Aoke ordered 56 picture books and 58 poetry books for the Wise Owl book store. How many books did he order altogether?

_____

27. Use the digits ③, ⑤, ⑨, and ④ only once to make 2 addends whose sum is between 100 and 110.

_____

Three-Digit Addition

Name_____

Date _____

Estimate. **Add.**

$$425 \longrightarrow 400$$
$$\underline{+188} \longrightarrow \underline{+200}$$
about 600

h	t	o
	1	
4	2	5
+1	8	8
		3

h	t	o
1	1	
4	2	5
+1	8	8
	1	3

h	t	o
1	1	
4	2	5
+1	8	8
6	1	3

13 ones =   11 tens =
1 ten 3 ones   1 hundred 1 ten

Complete.

1. 123
     +139
       2

2. 397
     + 45
       4

3. 132
     +184
       6

4. 576
     +136
       7

5. 273
     +472
       5

Estimate. Then add.

6. 257
     +136

7. 274
     +253

8. 398
     +375

9. 192
     +239

10. 388
      +162

11. 226
      +218

12. 196
      + 95

13. $3.54
      + 2.73

14. $1.46
      + 3.96

15. $4.87
      + 3.75

Find the sum.

16. 345
      +185

17. 627
      +297

18. 304
      +206

19. 366
      +359

20. 806
      + 84

21. 938 + 52 = _____

22. 46 + 265 = _____

23. 474 + 317 = _____

24. 731 + 169 = _____

PROBLEM SOLVING

25. Find the total of 446 first-class
stamps and 473 airmail stamps. _____

Mental Math

Name _____

Date _____

Break apart numbers to find tens.	Look for patterns.
$32 + 47 = \underline{?}$	You know $9 + 8 = 17$.
Think: $(30 + 2) + (40 + 7)$	So $19 + 8 = 27$.
$(30 + 40) + (2 + 7)$	$29 + 8 = 37$
$70 + 9 = 79$	$39 + 8 = 47$

Add mentally.

1. $16 + 6 =$ _____

2. $26 + 6 =$ _____

3. $36 + 6 =$ _____

4. $25 + 9 =$ _____

5. $35 + 9 =$ _____

6. $45 + 9 =$ _____

7. $57 + 8 =$ _____

8. $74 + 7 =$ _____

9. $93 + 8 =$ _____

10. $87 + 6 =$ _____

11. $13 + 9 =$ _____

12. $26 + 9 =$ _____

13. $64 + 15 =$ _____

14. $34 + 64 =$ _____

15. $42 + 34 =$ _____

16. $\begin{array}{r} 19 \\ + 32 \\ \hline \end{array}$

17. $\begin{array}{r} 44 \\ + 61 \\ \hline \end{array}$

18. $\begin{array}{r} 57 \\ + 25 \\ \hline \end{array}$

19. $\begin{array}{r} 71 \\ + 16 \\ \hline \end{array}$

20. $\begin{array}{r} 56 \\ + 61 \\ \hline \end{array}$

21. $\begin{array}{r} 84 \\ + 26 \\ \hline \end{array}$

22. $\begin{array}{r} 33 \\ + 38 \\ \hline \end{array}$

23. $\begin{array}{r} 66 \\ + 45 \\ \hline \end{array}$

24. $\begin{array}{r} 29 \\ + 49 \\ \hline \end{array}$

25. $\begin{array}{r} 46 \\ + 18 \\ \hline \end{array}$

PROBLEM SOLVING

26. Kathleen had 24 glass birds in her collection. For her birthday, she received 37 more glass birds. How many glass birds does Kathleen have now?

_____

Regrouping Hundreds

Name_____

Date _____

Regroup.

1 thousand 12 hundreds =
2 thousands 2 hundreds

23 hundreds =
2 thousands 3 hundreds

Regroup.

1. 15 hundreds =
_____ thousand _____ hundreds

2. 19 hundreds =
_____ thousand _____ hundreds

3. 12 hundreds =
_____ thousand _____ hundreds

4. 6 thousands 13 hundreds =
_____ thousands _____ hundreds

5. 1 thousand 18 hundreds =
_____ thousands _____ hundreds

6. 8 thousands 16 hundreds =
_____ thousands _____ hundreds

7. 3 thousands 22 hundreds =
_____ thousands _____ hundreds

8. 2 thousands 24 hundreds =
_____ thousands _____ hundreds

9. 3 thousands 34 hundreds =
_____ thousands _____ hundreds

10. 5 thousands 36 hundreds =
_____ thousands _____ hundreds

Match.

11. 6 thousands 38 hundreds = _____

a) 7 thousands 4 hundreds

12. 2 thousands 23 hundreds = _____

b) 8 thousands 6 hundreds

13. 4 thousands 18 hundreds = _____

c) 4 thousands 3 hundreds

14. 3 thousands 44 hundreds = _____

d) 9 thousands 8 hundreds

15. 7 thousands 16 hundreds = _____

e) 5 thousands 8 hundreds

16. 5 thousands 45 hundreds = _____

f) 9 thousands 5 hundreds

Use with Lesson 2–14 , text page 89.

27

Three or More Addends

Name _____

Date _____

Estimate.   **Add.**

$4.22 ⟶ $4.00
  4.54 ⟶   5.00
+ 2.36 ⟶ + 2.00
      about $11.00

```
    1               1  1            1  1
$4.2 2          $4.2 2          $4.2 2
 4.5 4           4.5 4           4.5 4
+ 2.3 6         + 2.3 6         + 2.3 6
      2              1 2        $11.1 2
```

Estimate. Then add.

1. 249	2. 421	3. 472	4. 739	5. 173
344	178	249	73	271
+423	+465	+256	+164	+ 18

6. 123	7. 217	8. $3.80	9. $4.92	10. $5.69
526	324	1.25	2.79	3.34
+136	+645	+ 5.55	+ 7.21	+ 9.53

Find the sum.

11. 123	12. 221	13. 233	14. 616	15. 429
247	419	344	117	772
423	320	602	201	533
+145	+180	+182	+383	+155

Grapes $1.15	Meat Pie $3.25	Bread $2.49	Apples $2.10	Oranges $3.85

Use the pictures to find the total cost for each.

16. oranges, apples, and grapes _____

17. bread, meat pie, and oranges _____

18. meat pie, grapes, and bread _____

19. oranges, grapes, and meat pie _____

20. bread, apples, and oranges _____

28   **Use with Lesson 2–15, text pages 90–91.**

Adding Four-Digit Numbers

Name_____

Date _____

Estimate.

5627 → 6000
+2394 → +2000
about 8000

Add.

th	h	t	o
		1	
5	6	2	7
+2	3	9	4
			1

th	h	t	o
	1	1	
5	6	2	7
+2	3	9	4
		2	1

th	h	t	o
	1	1	1
5	6	2	7
+2	3	9	4
	0	2	1

th	h	t	o
1	1	1	
5	6	2	7
+2	3	9	4
8	0	2	1

Estimate. Then add.

1. 3768
   +2986

2. 2492
   + 639

3. 1986
   +7123

4. 5345
   + 985

5. 5098
   + 109

6. 2985
   +1328

7. 1357
   +5456

8. 4930
   +2109

9. 2794
   +5739

10. 3850
    +2465

Find the sum. Use the $ and . when needed.

11. 2552
    + 331

12. 1678
    + 848

13. 4433
    +3348

14. $18.60
    +   5.62

15. $44.41
    +   .71

16. 2929
    +6363

17. 1829
    +3453

18. $33.32
    + 56.74

19. $24.21
    + 47.32

20. $45.37
    + 13.26

21. 2537 + 1236 = _____

22. 1483 + 8263 = _____

23. 3839 + 4375 = _____

24. 2648 + 5566 = _____

25. 2987 + 3979 = _____

26. 5986 + 3659 = _____

PROBLEM SOLVING

27. Monday morning, 2466 cars crossed
the bridge to the city. During the rest
of the day, 3254 more cars crossed
the bridge. How many cars
crossed the bridge on Monday?

_____

Use with Lesson 2–16 , text pages 92–93.

29

Problem-Solving Strategy: Use Simpler Numbers

Name _____

Date _____

Joe and Fred need 876 bricks to build a chimney.
They also need 327 more to build the fireplace.
How many bricks do they need?

$$
\begin{array}{r}
^{1\ 1} \\
876 \\
+327 \\
\hline
1203
\end{array}
$$

Use simpler numbers. Use 80 for 876 and use 30 for 327.
Add: $80 + 30 = 110$ Now compute the numbers in the problem.
Joe and Fred need to buy 1203 bricks.

Solve. Do your work on a separate sheet of paper.

1. Joe counted three bundles of boards. One had 276 boards, another had 319, and the third had 121. How many boards were in the three bundles?

2. Fred paid $75.50 for house plans, $9.75 for a hammer, and $13.50 for a saw. How much did these items cost altogether?

3. Each side of a house needed a different amount of siding. The front needed 248 feet and the back needed 288 feet. The sides needed 450 and 420 feet. How many feet were needed in all?

4. Fred wanted wood floors in the house. He needed 578 feet of wood flooring upstairs and 725 feet downstairs. What was the total number of feet that he needed?

5. Joe used 3 spools of wire for the electricity. Each spool had 250 feet. How many feet of wire were used?

6. The roof needed 140 shingles on one side and 185 on the other. How many shingles did Joe need to buy?

7. Joe insulated a house. The walls needed 565 yards of insulation and the roof needed 378 yards. How many yards did Joe have to buy?

8. A house that Joe and Fred were building called for 137 long studs and 98 short studs. How many studs did they need to buy?

Subtraction Concepts

Name_____

Date _____

Sam's dog had 5 puppies.
He gave away 3 puppies.
How many are left?

Take Away
$5 - 3 = 2$
Two puppies are left.

There are 5 puppies.
One puppy is brown.
How many are not brown?

Find Part of a Whole Set
$5 - 1 = 4$
Four are not brown.

Sam's dog had 5 puppies.
Grace's dog had 7 puppies.
How many more puppies did
Grace's dog have?

Compare
$7 - 5 = 2$
Grace's dog had 2 more puppies.

Grace has $6. She needs $9.
How many more dollars
does she need?

How Many More Are Needed
$\$9 - \$6 = \$3$
She needs $3 more.

PROBLEM SOLVING Write which meaning of subtraction you used.

1. The Pretty Pooch Pet Store had 19 beagles. They
   sold 4 beagles. How many beagles were left? _____

2. Jack collects animal stamps. He can fit 45
   stamps in an album. He has 32 stamps. How
   many more does he need to fill the album? _____

3. Miko has 22 tropical fish. Alex has 10 fish. How
   many more fish does Miko have? _____

4. Anna wants to buy a book about cats that
   costs $18. She has $9. How much more
   money does she need? _____

5. There are 26 kittens at the pet store. Five kittens
   are gray. How many kittens are not gray? _____

Subtracting: No Regrouping

Name_____

Date_____

Estimate.   **Subtract.**

Use front-end estimation.

283	→	200
−121	→	−100
		about 100

h	t	o
2	8	3
−1	2	1
		2

h	t	o
2	8	3
−1	2	1
	6	2

h	t	o
2	8	3
−1	2	1
1	6	2

Estimate. Then find the difference.

1. 55
−33

2. 48
−23

3. 29
−27

4. 62
−61

5. 65
−21

6. 86
−24

7. 16
− 5

8. 57
− 4

9. 95
−65

10. 71
−21

11. 57
−20

12. 89
−45

Subtract.

13. 289
−224

14. 742
−231

15. 768
−243

16. 356
− 51

17. 597
−185

18. 429
−115

19. 865
−832

20. 459
−245

21. 675
−371

22. 986
−543

Align and subtract.

23. 47 − 7

24. 58 − 28

25. 21 − 11

26. 234 − 21

27. 356 − 33

28. 658 − 232

Subtracting Money

Name_____

Date _____

Estimate. Then subtract.

1. $.6 7
    − .2 4

2. $.8 9
    − .7 8

3. $.9 8
    − .1 8

4. $.2 9
    − .0 4

5. $5.3 9
    − 5.0 2

6. $8.7 5
    − 8.1 3

7. $2.8 6
    − .2 0

8. $9.9 9
    − 6.5 5

9. $7.4 2
    − 1.4 1

10. $4.1 8
    − 3.1 2

11. $3.92
    − .81

12. $6.0 8
    − 2.0 7

13. $9.8 3
    − 1.1 3

14. $1.8 4
    − .0 4

15. $5.7 7
    − 4.6 7

16. $8.1 8
    − 3.0 5

PROBLEM SOLVING

17. Miranda bought a notebook for $3.39
    and a pen for $1.19. How much
    more did she pay for the notebook?

    _____

18. Walter bought a ruler for $.79 and an
    eraser for $.25. How much more
    did he pay for the ruler?

    _____

Estimating Differences

Name_____

Date _____

Round to the nearest:

Ten	Hundred	Ten Cents	Dollar
55 → 60 −24 → −20 40	838 → 800 −269 → −300 500	$.77 → $.80 − .23 → − .20 $.60	$18.75 → $19 − 5.12 → − 5 $14

Estimate by rounding to the nearest ten or ten cents.

1. 6 7
   −1 2

2. 6 5
   −3 2

3. 7 3
   −1 5

4. 6 0
   −4 8

5. $.2 5
   − .1 1

6. $2.5 9
   − .1 9

7. $.4 6
   − .3 1

8. $2.7 5
   − 1.2 8

Estimate by rounding to the nearest hundred or dollar.

9. 2 5 9
   −1 1 2

10. 5 9 8
   −1 9 9

11. 6 4 6 7
    −2 3 1 2

12. 3 8 8 5
    −1 3 1 2

13. $3.72 − $1.24 _____

14. $65.46 − $22.20 _____

PROBLEM SOLVING

15. During one hour at the airport, $876 was taken in from food sales. The next hour, $564 was collected. **About** how much less was collected during the second hour?   _____

Estimate how much less than $876 was collected using each amount:

16. $328 _____

17. $476 _____

18. $211 _____

19. $799 _____

20. $156 _____

21. $651 _____

Subtracting with Regrouping

Estimate by rounding.

44 ⟶ 40
−18 ⟶ −20
about 20

Subtract.
Regroup as needed.

tens	ones
3 4̸	14 4̸
− 1	8
	6

4 tens 4 ones = 3 tens 14 ones

tens	ones
3 4̸	14 4̸
− 1	8
2	6

Estimate. Then find the difference.

1. 85
−26

2. 92
−36

3. 65
−28

4. 43
−26

5. 72
−35

6. 56
−48

7. 24
−18

8. 47
−39

9. 83
−46

10. 74
−29

11. 56
− 8

12. 94
−38

Align and subtract.

13. 62 − 24

14. 70¢ − 52¢

15. $.68 − $.19

16. 45 − 36

17. 55¢ − 37¢

18. $.75 − $.36

19. $.72 − $.57

20. 63¢ − 29¢

21. 78 − 49

PROBLEM SOLVING

22. Randall had 42 stickers. He gave some away.
Then he had 27 left. How many did he give away? _____

Regrouping Hundreds and Dollars

Name_____

Date_____

6 hundreds 8 tens =	7 dollars 3 dimes =
5 hundreds + 1 hundred + 8 tens =	6 dollars + 1 dollar + 3 dimes =
5 hundreds + 10 tens + 8 tens =	6 dollars + 10 dimes + 3 dimes =
5 hundreds + 18 tens =	6 dollars + 13 dimes =
5 hundreds 18 tens	6 dollars 13 dimes

Regroup. You may use base ten blocks.

1. 7 hundreds 2 tens =

_____ hundreds _____tens

2. 3 hundreds 4 tens =

_____ hundreds _____tens

3. 2 hundreds 0 tens =

_____ hundred _____tens

4. 9 hundreds 3 tens =

_____ hundreds _____tens

5. 8 hundreds 8 tens =

_____ hundreds _____tens

6. 4 hundreds 5 tens =

_____ hundreds _____tens

7. 9 hundreds 1 ten =

_____ hundreds _____tens

8. 7 hundreds 7 tens =

_____ hundreds _____tens

9. 1 dollar 1 dime =

_____ dollars _____dimes

10. 6 dollars 1 dime =

_____ dollars _____dimes

11. 5 dollars 0 dimes =

_____ dollars _____dimes

12. 3 dollars 2 dimes =

_____ dollars _____dimes

13. 8 dollars 6 dimes =

_____ dollars _____dimes

14. 4 dollars 4 dimes =

_____ dollars _____dimes

15. 2 dollars 2 dimes =

_____ dollar _____dimes

16. 5 dollars 8 dimes =

_____ dollars _____dimes

17. 6 hundreds 7 tens = _____hundreds _____tens

Regrouping Once in Subtraction

Name_____

Date _____

Estimate by rounding.

328	$\longrightarrow$	300
−182	$\longrightarrow$	−200
	about 100	

Subtract. Regroup as needed.

h	t	o
3	2	8
−1	8	2
		6

h	t	o
²3̶	¹²2̶	8
−1	8	2
	4	6

h	t	o
²3̶	¹²2̶	8
−1	8	2
1	4	6

Complete.

1. 417
   −153
   64

2. 239
   −184

3. 927
   −435
   9

4. 179
   − 84

5. 878
   −296
   5

Estimate. Then subtract.

6. 765
   −273

7. 349
   −157

8. 567
   − 87

9. 454
   −162

10. 238
   − 65

11. 375
   −126

12. 945
   −372

13. 186
   − 93

14. 284
   −116

15. 416
   −107

16. $2.35
   − 1.41

17. $6.78
   − 4.29

18. $3.86
   − 2.94

19. $3.15
   − 1.22

20. $8.88
   − 7.95

Align and subtract.

21. 397 − 148

22. 276 − 81

23. 756 − 273

24. 476 − 388

25. 989 − 899

26. 678 − 592

27. $6.18 − $.31

28. $7.11 − $2.80

29. $5.25 − $1.31

30. $6.18 − $3.43

 37

Regrouping Twice in Subtraction

Name_____

Date _____

Estimate by rounding.

$$873 \longrightarrow 900$$
$$-587 \longrightarrow -600$$
$$\text{about } 300$$

Subtract.

h	t	o
8	7	3
5	8	7
		6

h	t	o
8	7	3
5	8	7
	8	6

h	t	o
8	7	3
5	8	7
2	8	6

Complete.

1.   7 5 0
     −3 9 3
        5 7

2.   3 2 2
     −2 2 8
          4

3.   8 5 4
     −2 7 6
        5   8

4.   7 6 3
     −1 8 5
          5

Estimate. Then subtract.

5.   5 6 7
     −1 9 8

6.   4 7 8
     −2 7 9

7.   5 8 1
     −2 9 6

8.   5 5 5
     −1 7 6

9.   9 1 1
     −4 8 3

10.   9 3 1
      −3 7 5

11.   4 2 3
      −1 2 8

12. $6.7 3
      − 4.8 7

13. $4.7 8
      − 3.9 9

14. $6.1 7
      − 1.2 8

Align and subtract.

15. 715 − 236

16. 853 − 284

17. 691 − 398

18. $7.88 − $2.89

19. $9.64 − $3.86

PROBLEM SOLVING

20. How much change will Tom receive if he spends $3.47 and gives the clerk a five-dollar bill and 2 quarters? _____

Regrouping with Zeros

Name_____

Date _____

Subtract. Regroup as needed.

h	t	o
⁴5̶	¹⁰0̶	0
− 3	7	9

h	t	o
⁴5̶	⁹⁄¹⁰0̶	¹⁰0̶
− 3	7	9

h	t	o
⁴5̶	⁹⁄¹⁰0̶	¹⁰0̶
− 3	7	9
1	2	1

Estimate. Then subtract.

1. 4 0 0
     − 1 1 8

2. 3 0 0
     − 2 2 5

3. 7 0 0
     − 3 4 6

4. 3 0 0
     − 1 0 7

5. 7 0 0
     − 1 2 8

6. 5 0 0
     − 2 5 7

7. 8 0 0
     − 4 2 4

8. $6.0 0
     − 3.0 5

9. $4.0 0
     − 2.7 7

10. $3.0 0
     − 1.9 8

11. $1.0 0
     −   .7 5

12. $2.0 0
     −   .2 8

13. $9.0 0
     − 5.1 9

14. $4.0 0
     − 1.1 8

15. $3.0 0
     − 2.2 5

16. $7.0 0
     − 5.5 5

Align and subtract.

17. 900 − 575

18. 500 − 275

19. $6.00 − $1.49

20. 206 − 118

21. 808 − 239

22. $5.02 − $2.23

PROBLEM SOLVING

23. The video store has 800 movies to rent. One weekend, all but 298 movies were rented. How many movies were rented that weekend?

_____

Choose a Computation Method

Name _____

Date _____

Subtract: 675 − 300 = __?__

- Can you subtract mentally? Use mental math.
- Are the numbers too large to subtract mentally? Will it take too much time to use a calculator? Use paper and pencil.
- Are there too many steps? Is a calculator handy? Use a calculator.

$$\begin{array}{r} 675 \\ -\ 300 \\ \hline 375 \end{array}$$

Subtract. Use mental math, paper and pencil, or a calculator.

1. $\begin{array}{r} 86 \\ -20 \\ \hline \end{array}$
2. $\begin{array}{r} 90 \\ -60 \\ \hline \end{array}$
3. $\begin{array}{r} 42 \\ -27 \\ \hline \end{array}$
4. $\begin{array}{r} 50 \\ -38 \\ \hline \end{array}$
5. $\begin{array}{r} \$.99 \\ -\ .10 \\ \hline \end{array}$
6. $\begin{array}{r} \$.62 \\ -\ .45 \\ \hline \end{array}$

7. $\begin{array}{r} 250 \\ -200 \\ \hline \end{array}$
8. $\begin{array}{r} 600 \\ -457 \\ \hline \end{array}$
9. $\begin{array}{r} \$4.95 \\ -\ 3.50 \\ \hline \end{array}$
10. $\begin{array}{r} \$2.38 \\ -\ .38 \\ \hline \end{array}$
11. $\begin{array}{r} \$8.50 \\ -\ 5.26 \\ \hline \end{array}$

12. $\begin{array}{r} 913 \\ -285 \\ \hline \end{array}$
13. $\begin{array}{r} 400 \\ -100 \\ \hline \end{array}$
14. $\begin{array}{r} 605 \\ -405 \\ \hline \end{array}$
15. $\begin{array}{r} \$9.77 \\ -\ 5.01 \\ \hline \end{array}$
16. $\begin{array}{r} \$1.75 \\ -\ .25 \\ \hline \end{array}$

17. $\begin{array}{r} 725 \\ -488 \\ \hline \end{array}$
18. $\begin{array}{r} 250 \\ -150 \\ \hline \end{array}$
19. $\begin{array}{r} 832 \\ -830 \\ \hline \end{array}$
20. $\begin{array}{r} \$5.63 \\ -\ 5.63 \\ \hline \end{array}$
21. $\begin{array}{r} \$4.27 \\ -\ 2.93 \\ \hline \end{array}$

22. Which exercises were you able to do mentally?

_____

23. Which exercises did you do using paper and pencil?

_____

Regrouping Thousands as Hundreds

Name_____

Date_____

> 2 thousands 4 hundreds = 1 thousand + 1 thousand + 4 hundreds
> = 1 thousand + 10 hundreds + 4 hundreds
> = 1 thousand + 14 hundreds
> = 1 thousand 14 hundreds

Regroup.

1. 3 thousands 7 hundreds =

_____thousands _____hundreds

2. 6 thousands 4 hundreds =

_____thousands _____hundreds

3. 1 thousand 6 hundreds =

_____thousands _____hundreds

4. 5 thousands 5 hundreds =

_____thousands _____hundreds

5. 8 thousands 1 hundred =

_____thousands _____hundreds

6. 3 thousands 2 hundreds =

_____thousands _____hundreds

7. 9 thousands 6 hundreds =

_____thousands _____hundreds

8. 7 thousands =

_____thousands _____hundreds

9. 4 thousands =

_____thousands _____hundreds

10. 3 thousands 2 hundreds =

_____thousands _____hundreds

11. 2 thousands 8 hundreds =

_____thousand _____hundreds

12. 8 thousands =

_____thousands _____hundreds

13. 7 thousands 6 hundreds =

_____thousands _____hundreds

14. 6 thousands 2 hundreds =

_____thousands _____hundreds

Use with Lesson 3–11, text page 121.
41

Subtracting Larger Numbers

Subtract: 7022 − 5974 = _?_

th	h	t	o
7	0	2̶ ¹	2̶ ¹²
− 5	9	7	4
			8

th	h	t	o
7̶ ⁶	0̶ ⁹ ¹⁰	2̶ ¹	2̶ ¹²
− 5	9	7	4
			8

th	h	t	o
7̶ ⁶	0̶ ⁹ ¹⁰	2̶ ¹	2̶ ¹²
− 5	9	7	4
	0	4	8

th	h	t	o
7̶ ⁶	0̶ ⁹ ¹⁰	2̶ ¹	2̶ ¹²
− 5	9	7	4
1	0	4	8

Check
```
  1048
+ 5974
  7022
```

Estimate. Then subtract.

1. 5762
    −1814

2. 3057
    − 832

3. 7582
    −2816

4. 6005
    −2316

5. 8000
    −2143

6. 2876
    − 951

7. 5000
    −1638

8. 8736
    −6495

9. $20.00
    −   4.53

10. $63.95
    − 28.94

11. $76.85
    − 38.69

12. $50.00
    − 28.59

13. $78.95
    − 34.26

Subtract. Then check by addition.

Check

14. 3789
    −1245

15. 5010
    −2090

Check

16. 4201
    −1029

Check

Check

17. 7543
    −5126

18. $89.56
    − 37.28

Check

19. $59.57
    − 32.84

Check

PROBLEM SOLVING

20. The Fresh Fruit Market is getting a shipment of 9000 oranges and 4250 apples. How many more oranges than apples are in the shipment?

_____

Problem-Solving Strategy: Choose the Operation

Name_____

Date _____

Jed bought a window box for $12.75, a flower pot for $7.95 and a hanging planter for $16.30. How much did Jed spend?

Jed spent $37.00

Think: Add $12.75 + $7.95 + $16.30

$$\begin{array}{r} \$12.75 \\ 7.95 \\ + \ 16.30 \\ \hline \$37.00 \end{array}$$

Solve. Do your work on a separate sheet of paper.

1. The garden store's bulb bins contain 237 tulip bulbs, 466 daffodil bulbs and 719 crocus bulbs. How many bulbs are there in all?

_____

2. Carroll wants to fence in his flower bed. He needs 1246 feet of fencing. The store has 1098 feet in stock. How much more fencing does Carroll need?

_____

3. At the end of the week, there were 485 3-inch peat pots left in the store. At the beginning of the week, there had been 2390 pots. How many pots were sold that week?

_____

4. Arnie wants to plant a new lawn. A grass rake costs $17.44, hay mulch costs $14.60, and 25 pounds of grass seed costs $45.19. How much will it cost Arnie to plant a lawn?

_____

5. Aretha worked 32 hours, 30 hours, and 29 hours during three weeks in June. How many hours did she work altogether?

_____

6. Electric hedge clippers cost $33.72. A rotary lawn mower costs $85.95. How much more expensive is the rotary mower?

_____

7. Evita buys a window box for $10.99. She also buys plants that cost $14.50. How much does she spend?

_____

8. Luis sold 958 packets of seeds. Jack sold 1322 packets. How many more packets did Jack sell than Luis?

_____

      43

Understanding Multiplication

Name _____

Date _____

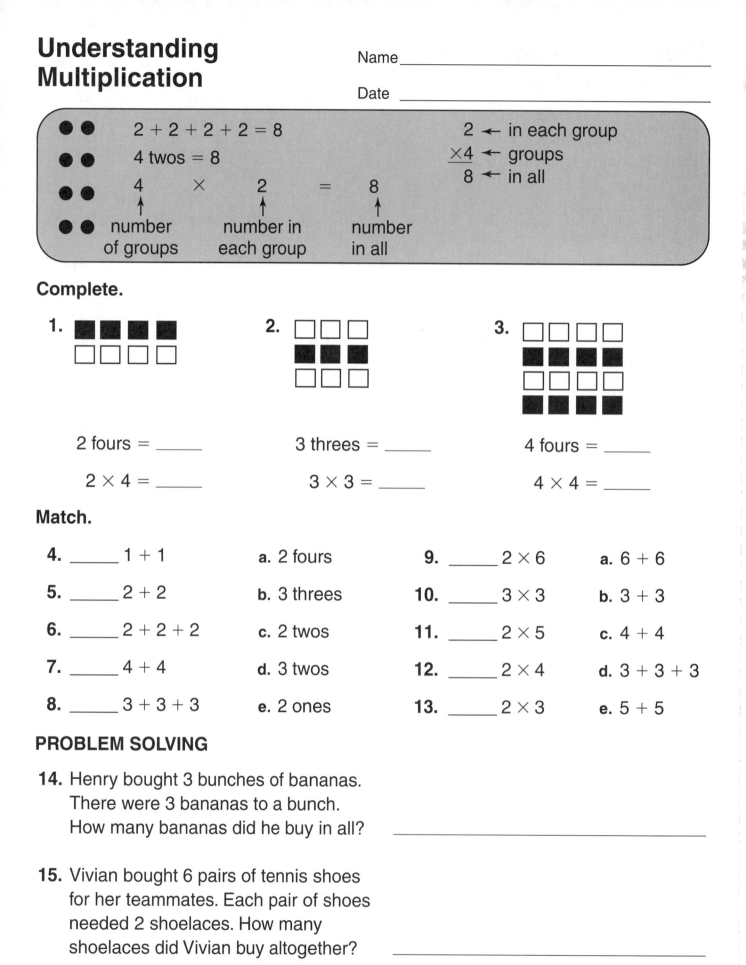

2 + 2 + 2 + 2 = 8

4 twos = 8

4 × 2 = 8

↑ number of groups   ↑ number in each group   ↑ number in all

2 ← in each group
×4 ← groups
8 ← in all

Complete.

1. ■ ■ ■ ■
□ □ □ □

2 fours = _____

2 × 4 = _____

2. □ □ □
■ ■ ■
□ □ □

3 threes = _____

3 × 3 = _____

3. □ □ □ □
■ ■ ■ ■
□ □ □ □
■ ■ ■ ■

4 fours = _____

4 × 4 = _____

Match.

4. _____ 1 + 1   **a.** 2 fours

5. _____ 2 + 2   **b.** 3 threes

6. _____ 2 + 2 + 2   **c.** 2 twos

7. _____ 4 + 4   **d.** 3 twos

8. _____ 3 + 3 + 3   **e.** 2 ones

9. _____ 2 × 6   **a.** 6 + 6

10. _____ 3 × 3   **b.** 3 + 3

11. _____ 2 × 5   **c.** 4 + 4

12. _____ 2 × 4   **d.** 3 + 3 + 3

13. _____ 2 × 3   **e.** 5 + 5

PROBLEM SOLVING

14. Henry bought 3 bunches of bananas. There were 3 bananas to a bunch. How many bananas did he buy in all? _____

15. Vivian bought 6 pairs of tennis shoes for her teammates. Each pair of shoes needed 2 shoelaces. How many shoelaces did Vivian buy altogether? _____

Zero and One as Factors

Name_____

Date _____

$$2 \times 1 = 2 \quad \textbf{or} \quad \begin{array}{r} 1 \leftarrow \text{factor} \\ \times 2 \leftarrow \text{factor} \\ \hline 2 \leftarrow \text{product} \end{array} \quad \Big| \quad 3 \times 0 = 0 \quad \textbf{or} \quad \begin{array}{r} 0 \\ \times 3 \\ \hline 0 \end{array}$$

factor factor product

Multiply.

1. $7 \times 1 =$ _____

2. $4 \times 1 =$ _____

3. $6 \times 0 =$ _____

4. $9 \times 0 =$ _____

5. $8 \times 1 =$ _____

6. $7 \times 0 =$ _____

7. $3 \times 1 =$ _____

8. $4 \times 0 =$ _____

9. $2 \times 0 =$ _____

Complete.

10.
$$\begin{array}{r} \times 1 \\ \hline 0 \end{array}$$

11.
$$\begin{array}{r} 1 \\ \times \\ \hline 1 \end{array}$$

12.
$$\begin{array}{r} 1 \\ \times 2 \\ \hline \end{array}$$

13.
$$\begin{array}{r} \times 3 \\ \hline 3 \end{array}$$

14.
$$\begin{array}{r} 1 \\ \times 4 \\ \hline \end{array}$$

15.
$$\begin{array}{r} 1 \\ \times \\ \hline 5 \end{array}$$

16.
$$\begin{array}{r} \times 2 \\ \hline 0 \end{array}$$

17.
$$\begin{array}{r} \times 3 \\ \hline 0 \end{array}$$

18.
$$\begin{array}{r} 0 \\ \times 4 \\ \hline \end{array}$$

19.
$$\begin{array}{r} 1 \\ \times 5 \\ \hline \end{array}$$

20.
$$\begin{array}{r} 1 \\ \times 6 \\ \hline \end{array}$$

21.
$$\begin{array}{r} \times 7 \\ \hline 0 \end{array}$$

PROBLEM SOLVING

22. My product is 5. One of my factors is 5. What is my other factor?

_____

23. One of my factors is 8. My product is 0. What is my other factor?

_____

24. Rosalie sent 1 postcard to each of her 7 friends. How many postcards did she send?

_____

Multiplying Twos

5 groups of 2	5 × 2 = 10	*or*	2 ← factor
	↑ ↑ ↑		× 5 ← factor
	factor factor product		10 ← product

Complete.

1.

_____ twos = 6

3 × 2 = _____

2.

_____ twos = 4

2 × 2 = _____

3.

_____ twos = 12

6 × 2 = _____

Multiply.

4.   2
      × 4

5.   2
      × 8

6.   2
      × 7

Find the product.

7. 1 × 2 = _____   **8.** 5 × 2 = _____   **9.** 8 × 2 = _____   **10.** 4 × 2 = _____

11.   2
       × 3

12.   2
       × 6

13.   2
       × 9

14.   2
       × 7

15.   2
       × 5

16.   2
       × 2

PROBLEM SOLVING

17. Ms. Johnson bought 9 packages of sponges. Each package contained 2 sponges. How many sponges did she buy?

_____

18. The factors are 7 and 2. What is the product?

_____

Multiplying Threes

Name_____

Date_____

4 groups of 3

$4 \times 3 = 12$ **or**   3 ← factor
$\times 4$ ← factor

factor factor product   12 ← product

Complete.

1.
_____ threes = 21

$7 \times 3 =$ _____

2.
_____ threes = 15

$5 \times 3 =$ _____

3.
_____ threes = 18

$6 \times 3 =$ _____

Multiply.

4.
$\begin{array}{r} 3 \\ \times 2 \\ \hline \end{array}$

5.
$\begin{array}{r} 3 \\ \times 5 \\ \hline \end{array}$

6.
$\begin{array}{r} 3 \\ \times 4 \\ \hline \end{array}$

Find the product.

7. $8 \times 3 =$ _____   **8.** $2 \times 3 =$ _____   **9.** $9 \times 3 =$ _____   **10.** $4 \times 3 =$ _____

11. $\begin{array}{r} 3 \\ \times 7 \\ \hline \end{array}$ **12.** $\begin{array}{r} 3 \\ \times 9 \\ \hline \end{array}$ **13.** $\begin{array}{r} 3 \\ \times 1 \\ \hline \end{array}$ **14.** $\begin{array}{r} 3 \\ \times 8 \\ \hline \end{array}$ **15.** $\begin{array}{r} 3 \\ \times 3 \\ \hline \end{array}$ **16.** $\begin{array}{r} 3 \\ \times 6 \\ \hline \end{array}$

PROBLEM SOLVING

17. Jamal sold cards to raise money for the school band. Nine people each bought 3 boxes of cards. How many boxes did he sell?

_____

18. Chris won 2 sets of books for selling the most cards. Each set had 3 books. How many books did he win altogether?

_____

Multiplying Fours

4 groups of 4   4 × 4 = 16   *or* $\begin{array}{r} 4 \\ \times 4 \\ \hline 16 \end{array}$

Complete.

1.

_____ fours = 24

6 × 4 = _____

2.

_____ fours = 12

3 × 4 = _____

3.

_____ fours = 20

5 × 4 = _____

Multiply.

4. $\begin{array}{r} 4 \\ \times 8 \\ \hline \end{array}$

5. $\begin{array}{r} 4 \\ \times 5 \\ \hline \end{array}$

6. $\begin{array}{r} 4 \\ \times 2 \\ \hline \end{array}$

7. $\begin{array}{r} 4 \\ \times 9 \\ \hline \end{array}$

8. $\begin{array}{r} 4 \\ \times 1 \\ \hline \end{array}$

9. $\begin{array}{r} 4 \\ \times 7 \\ \hline \end{array}$

10. 9 × 4 = _____   11. 3 × 4 = _____   12. 7 × 4 = _____   13. 8 × 4 = _____

14. 6 × 4 = _____   15. 1 × 4 = _____   16. 4 × 4 = _____   17. 5 × 4 = _____

18. 8 × 4 = _____   19. 0 × 4 = _____   20. 9 × 4 = _____   21. 2 × 4 = _____

PROBLEM SOLVING

22. Lori made 3 salads. Each salad had 4 tomatoes. How many tomatoes did she use?

_____

23. Carlos made 8 sandwiches. Each sandwich had 4 slices of turkey. How many turkey slices did he use?

_____

Multiplying Fives

Name _____

Date _____

4 groups of 5 $4 \times 5 = 20$ *or* $\begin{array}{r} 5 \\ \times 4 \\ \hline 20 \end{array}$

Complete.

1.

2 fives = _____

$2 \times$ _____ $= 10$

2.

$1 \times$ _____ $= 5$

$1 \times 5 =$ _____

3.

$5 \times$ _____ $= 25$

_____ $\times$ _____ $= 25$

Multiply.

4.

$\begin{array}{r} 5 \\ \times 4 \\ \hline \end{array}$

5.

$\begin{array}{r} 5 \\ \times 8 \\ \hline \end{array}$

Find the product.

6. $1 \times 5 =$ _____

7. $8 \times 5 =$ _____

8. $9 \times 5 =$ _____

9. $3 \times 5 =$ _____

10. $7 \times 5 =$ _____

11. $6 \times 5 =$ _____

12. $4 \times 5 =$ _____

13. $2 \times 5 =$ _____

14. $\begin{array}{r} 5 \\ \times 2 \\ \hline \end{array}$

15. $\begin{array}{r} 5 \\ \times 7 \\ \hline \end{array}$

16. $\begin{array}{r} 5 \\ \times 5 \\ \hline \end{array}$

17. $\begin{array}{r} 5 \\ \times 6 \\ \hline \end{array}$

18. $\begin{array}{r} 5 \\ \times 3 \\ \hline \end{array}$

19. $\begin{array}{r} 5 \\ \times 9 \\ \hline \end{array}$

PROBLEM SOLVING

20. Michael made party bags for 4 friends.
He put 5 sticks of gum in each bag.
How many sticks of gum did he use?   _____

21. Michael received 2 sets of markers at his
birthday party. Each set had 5 markers.
How many markers did he receive altogether?   _____

Use with Lesson 4–6, text pages 144–145.   Copyright © William H. Sadlier, Inc. All rights reserved.   49

Multiplying Cents

Name_____

Date _____

3¢ ← cents in each group
× 2 ← number of groups
6¢ ← cents in all

Multiply. Then write the ¢ sign when needed.

1.
$$4¢ \times 3$$

2.
$$2¢ \times 4$$

3. $2¢ \times 8$
4. $3¢ \times 1$
5. $5¢ \times 5$
6. $2¢ \times 6$
7. $1¢ \times 8$
8. $4¢ \times 5$

9. $2¢ \times 7$
10. $4¢ \times 8$
11. $1¢ \times 2$
12. $4¢ \times 7$
13. $4¢ \times 0$
14. $5¢ \times 3$

15. 5×7
16. $3¢ \times 8$
17. 2×9
18. $4¢ \times 9$
19. $5¢ \times 7$
20. 1×6

21. $4 \times 4¢ = $ _____

22. $2 \times 1 = $ _____

23. $8 \times 5 = $ _____

24. $4 \times 0 = $ _____

25. $5 \times 2¢ = $ _____

26. $7 \times 3 = $ _____

PROBLEM SOLVING

27. Pat bought 4 toy charms. Each charm cost 5¢. How much money did she spend?

_____

28. Charles bought 6 stickers. Each sticker cost 4¢. How much money did he spend?

_____

Sums, Differences, and Products

Name_____

Date _____

Add.	Subtract.	Multiply.
1 1	8 10 6 15	
1476	9̶0̶7̶5̶	3
+ 6059	− 1827	× 5
7535 ←sum	7248 ←difference	15 ← product

Add, subtract, or multiply. Watch the signs.

1. 428
    + 36

2. 8094
    +1093

3. 3¢
    × 9

4. 1
    × 9

5. 6035
    − 2508

6. 2
    × 5

7. 5
    × 6

8. 826
    + 407

9. 3672
    − 489

10. 0
     × 8

11. 43
     +19

12. 4¢
     × 2

13. 482
     −270

14. 596
     +324

15. $31.67
     + 42.85

16. $.81
     − .65

PROBLEM SOLVING

17. An adult has 206 bones. A newborn baby has 300 bones. How many more bones does a baby have?

    _____

18. The spine has 26 bones. The chest has 25 bones. How many bones are in the chest and spine altogether?

    _____

19. There are 5 bones in each palm. How many bones are in two palms?

    _____

Order in Multiplication

$$4 \times 3 = 12$$

number of groups   number in each group   number in all

$$3 \times 4 = 12$$

number of groups   number in each group   number in all

Find the product.

1. $8 \times 5 =$ _____

$5 \times 8 =$ _____

2. $7 \times 4 =$ _____

$4 \times 7 =$ _____

3. $9 \times 3 =$ _____

$3 \times 9 =$ _____

4. $6 \times 3 =$ _____

$3 \times 6 =$ _____

5. $6 \times 2 =$ _____

$2 \times 6 =$ _____

6. $5 \times 3 =$ _____

$3 \times 5 =$ _____

7. $9 \times 0 =$ _____

$0 \times 9 =$ _____

8. $7 \times 1 =$ _____

$1 \times 7 =$ _____

Multiply.

9.
$$\begin{array}{r} 4 \\ \times 5 \end{array} \qquad \begin{array}{r} 5 \\ \times 4 \end{array}$$

10.
$$\begin{array}{r} 3 \\ \times 7 \end{array} \qquad \begin{array}{r} 7 \\ \times 3 \end{array}$$

11.
$$\begin{array}{r} 5 \\ \times 2 \end{array} \qquad \begin{array}{r} 2 \\ \times 5 \end{array}$$

12.
$$\begin{array}{r} 7 \\ \times 5 \end{array} \qquad \begin{array}{r} 5 \\ \times 7 \end{array}$$

13.
$$\begin{array}{r} 6 \\ \times 4 \end{array} \qquad \begin{array}{r} 4 \\ \times 6 \end{array}$$

14.
$$\begin{array}{r} 9 \\ \times 4 \end{array} \qquad \begin{array}{r} 4 \\ \times 9 \end{array}$$

15. $4 \times 8 =$ ____ $\times 4$

16. $9 \times 2 = 2 \times$ ____

17. ____ $\times 6 =$ ____ $\times 1$

18. $3 \times$ ____ $= 8 \times$ ____

19. $7 \times 2 =$ ____ $\times$ ____

20. $6 \times 5 =$ ____ $\times$ ____

PROBLEM SOLVING

21. Anne put 2 labels on each of her video cassettes. She owns 9 video cassettes. How many labels did she use?

_____

22. Manuel made 2 cakes for his brother's birthday party. He used 3 eggs in each. How many eggs did he use in all?

_____

Missing Factors

Name_____

Date_____

$$\underline{\;?\;} \times 3 = 12 \longrightarrow 4 \times 3 = 12$$

factors   product   4 is the missing factor.

Find the missing factor.

1. $\begin{array}{r} 2 \\ \times\underline{} \\ \hline 12 \end{array}$

2. $\begin{array}{r} \times 3 \\ \hline 15 \end{array}$

3. $\begin{array}{r} \times 8 \\ \hline 32 \end{array}$

4. $\begin{array}{r} \times 8 \\ \hline 16 \end{array}$

5. $\begin{array}{r} \times 5 \\ \hline 20 \end{array}$

6. $\begin{array}{r} \times 3 \\ \hline 12 \end{array}$

7. $\begin{array}{r} \times 6 \\ \hline 24 \end{array}$

8. $\begin{array}{r} \times 7 \\ \hline 35 \end{array}$

9. $\begin{array}{r} \times 7 \\ \hline 14 \end{array}$

10. $\begin{array}{r} \times 3 \\ \hline 9 \end{array}$

11. $\begin{array}{r} \times 2 \\ \hline 6 \end{array}$

12. $\begin{array}{r} \times 7 \\ \hline 21 \end{array}$

13. $2 \times \underline{} = 8$

14. $\underline{} \times 5 = 40$

15. $7 \times \underline{} = 14$

16. $\underline{} \times 5 = 20$

17. $\underline{} \times 5 = 45$

18. $4 \times \underline{} = 16$

19. $18 = 3 \times \underline{}$

20. $9 = 9 \times \underline{}$

21. $28 = 7 \times \underline{}$

Complete.

22.

factor	factor	product
8		24
	6	18
5		20

23.

factor	factor	product
9		27
	5	45
8		32

PROBLEM SOLVING

24. Wendy put 16 carrots in bunches. She put 4 carrots in each bunch. How many bunches did she have? _____

25. Dave sewed 14 sails. He put the sails on 7 boats. Each boat had the same number of sails. How many did he put on each boat? _____

       53

Problem-Solving Strategy: Multi-Step Problems

Name_____

Date _____

> There were 77 third graders in three rooms. There were 27 third graders in Room 101 and 25 in Room 102. How many third graders were in Room 103?
>
> First find the number of third graders in Rooms 101 and 103.
> Add: 27 + 25 = 52
> Then find out how many are not in those rooms.
> Subtract: 77 − 52 = 25
> There were 25 third graders in Room 103.

Solve. Do your work on a separate sheet of paper.

1. Of the 77 third graders, 3 were absent from Room 101 on Monday, 4 were absent from Room 102, and 2 were absent from Room 103. How many third graders attended school that day?

   _____

2. Ms. Diaz gave 5 toothpicks to each of 9 children for an art project. The full box she started with held 100 toothpicks. How many toothpicks did she have left?

   _____

3. Ms. Diaz bought glue for $1.49 and poster paper for $4.50. How much change did she receive if she paid with a ten-dollar bill?

   _____

4. Mr. Vincent had a package of 35 pencils. He gave 2 pencils to each of 9 children. How many pencils did he have left?

   _____

5. At lunch, Alison spent $.25 for milk and $.35 for an orange. She gave the cashier $1.00. What was her change?

   _____

6. Mr. Wilbur gave 2 animal books to each of 5 children and 3 puzzle books to each of 4 children. How many books did he give to all the children?

   _____

Understanding Division

Name

Date

How many groups?		How many in each group?
12 baseballs in all		12 baseballs in all
3 baseballs in each group		4 groups of baseballs
? groups of baseball		? baseballs in each group
12 ÷ 3 = 4		12 ÷ 4 = 3
4 groups of baseballs		3 baseballs in each group

Complete.

1. ____ dots in all

____ groups of 2 dots

2. ____ dots in all

____ groups of 3 dots

3. ____ dots in all

____ groups of 5 dots

Write how many groups.

4. 10 dots in all
2 in each group

10 ÷ 2 = ____ groups

5. 12 dots in all
4 in each group

12 ÷ 4 = ____ groups

6. 15 dots in all
3 in each group

15 ÷ 3 = ____ groups

Write how many in each group.

7. 4 dots in all

2 groups of ____ dots

4 ÷ 2 = ____

8. 12 dots in all

2 groups of ____ dots

12 ÷ 2 = ____

9. 16 dots in all

4 groups of ____ dots

16 ÷ 4 = ____

Relating Multiplication and Division

Name _____

Date _____

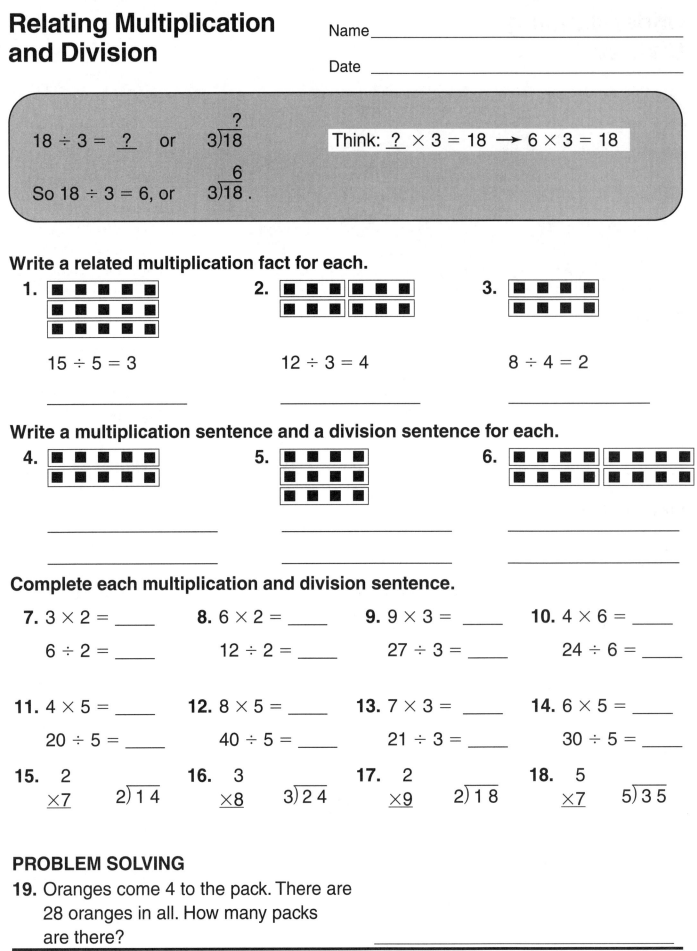

$18 \div 3 =$ __?__   or $3\overline{)18}^{?}$ Think: __?__ $\times 3 = 18 \longrightarrow 6 \times 3 = 18$

So $18 \div 3 = 6$, or $3\overline{)18}^{6}$.

Write a related multiplication fact for each.

1.

$15 \div 5 = 3$

_____

2.

$12 \div 3 = 4$

_____

3.

$8 \div 4 = 2$

_____

Write a multiplication sentence and a division sentence for each.

4.

_____

_____

5.

_____

_____

6.

_____

_____

Complete each multiplication and division sentence.

7. $3 \times 2 =$ _____

$6 \div 2 =$ _____

8. $6 \times 2 =$ _____

$12 \div 2 =$ _____

9. $9 \times 3 =$ _____

$27 \div 3 =$ _____

10. $4 \times 6 =$ _____

$24 \div 6 =$ _____

11. $4 \times 5 =$ _____

$20 \div 5 =$ _____

12. $8 \times 5 =$ _____

$40 \div 5 =$ _____

13. $7 \times 3 =$ _____

$21 \div 3 =$ _____

14. $6 \times 5 =$ _____

$30 \div 5 =$ _____

15. $\begin{array}{r} 2 \\ \times 7 \\ \hline \end{array}$ $2\overline{)14}$

16. $\begin{array}{r} 3 \\ \times 8 \\ \hline \end{array}$ $3\overline{)24}$

17. $\begin{array}{r} 2 \\ \times 9 \\ \hline \end{array}$ $2\overline{)18}$

18. $\begin{array}{r} 5 \\ \times 7 \\ \hline \end{array}$ $5\overline{)35}$

PROBLEM SOLVING

19. Oranges come 4 to the pack. There are 28 oranges in all. How many packs are there?

_____

Zero and One in Division

Name _____

Date _____

dividend	divisor	quotient		
0	÷ 2	= 0	or	0 ← quotient 2)0 ← dividend
				↑ divisor
4	÷ 1	= 4	or	4 4)1
3	÷ 3	= 1	or	1 3)3 .

Write the quotient. You may use counters.

1. 6 ÷ 1 = _____

2. 0 ÷ 5 = _____

3. 2 ÷ 2 = _____

4. 0 ÷ 3 = _____

5. 4 ÷ 4 = _____

6. 7 ÷ 1 = _____

7. 1 ÷ 1 = _____

8. 8 ÷ 8 = _____

9. 0 ÷ 9 = _____

10. 7 ÷ 7 = _____

11. 0 ÷ 6 = _____

12. 5 ÷ 1 = _____

13. 3 ÷ 1 = _____

14. 8 ÷ 1 = _____

15. 0 ÷ 2 = _____

16. 0 ÷ 4 = _____

17. 9 ÷ 9 = _____

18. 6 ÷ 6 = _____

19. 7)0   20. 5)5   21. 1)2   22. 8)8   23. 1)9   24. 8)0

PROBLEM SOLVING

25. Five friends shared 5 apples equally. How many apples did each friend get?

_____

26. There are 4 pears in all. How many pears are in 1 group?

_____

  57

Dividing by 2

Name_____

Date _____

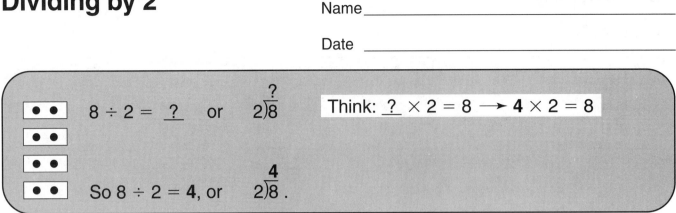

Write a division sentence for each model.

1. _____

2. _____

3. _____

Write the quotient. Skip count or use counters to help.

4. 2 ÷ 2 = _____

5. 10 ÷ 2 = _____

6. 14 ÷ 2 = _____

7. 18 ÷ 2 = _____

8. 8 ÷ 2 = _____

9. 16 ÷ 2 = _____

10. 0 ÷ 2 = _____

11. 4 ÷ 2 = _____

12. 12 ÷ 2 = _____

13. 2)6

14. 2)18

15. 2)10

16. 2)8

17. 2)4

18. 2)16

19. 2)2

20. 2)14

21. 2)0

22. 2)12

PROBLEM SOLVING

23. Shauna sold 16 boxes of cookies. She sold 2 boxes to each customer. How many customers did she have? _____

24. Jason has 18 white socks. How many pairs of white socks does he have? _____

Dividing By 3

Name _____

Date _____

$6 \div 3 =$ _?_ or $3\overline{)6}^{?}$ Think: _?_ $\times 3 = 6 \longrightarrow 2 \times 3 = 6$

So $6 \div 3 = 2$, or $3\overline{)6}^{2}$.

Write a division sentence for each model.

1. ⭐⭐⭐ / ⭐⭐⭐

2. ⭐⭐⭐ / ⭐⭐⭐ / ⭐⭐⭐ / ⭐⭐⭐

3. ⭐⭐⭐ ⭐⭐⭐ / ⭐⭐⭐ ⭐⭐⭐ / ⭐⭐⭐ / ⭐⭐⭐

_____   _____   _____

Write the quotient. Skip count or use counters to help.

4. $3\overline{)27}$ **5.** $3\overline{)9}$ **6.** $3\overline{)15}$ **7.** $3\overline{)24}$ **8.** $3\overline{)6}$

9. $3\overline{)3}$ **10.** $3\overline{)12}$ **11.** $3\overline{)21}$ **12.** $3\overline{)18}$ **13.** $3\overline{)0}$

14. $24 \div 3 =$ ____   **15.** $27 \div 3 =$ ____   **16.** $0 \div 3 =$ ____

17. $15 \div 3 =$ ____   **18.** $21 \div 3 =$ ____   **19.** $6 \div 3 =$ ____

20. $3 \div 3 =$ ____   **21.** $18 \div 3 =$ ____   **22.** $9 \div 3 =$ ____

PROBLEM SOLVING

23. Jason has 6 toy airplanes. He puts them in 3 equal groups. How many toy airplanes are in each group? _____

24. Wanda has 12 stuffed bears. She puts 3 bears on each shelf. How many shelves does she use? _____

Use with Lesson 5–5, text pages 172–173.   59

Dividing By 4

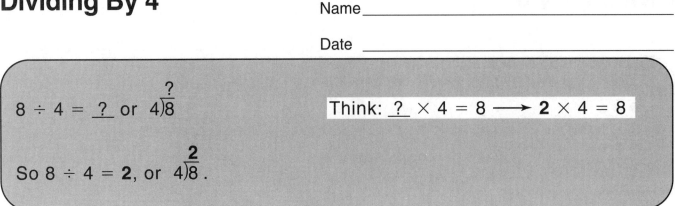

$8 \div 4 = \underline{\ ?\ }$ or $4\overline{)8}$

Think: $\underline{\ ?\ } \times 4 = 8 \longrightarrow 2 \times 4 = 8$

So $8 \div 4 = \mathbf{2}$, or $4\overline{)8}$.

Write a division sentence for each model.

1. ✰✰✰✰
✰✰✰✰
✰✰✰✰

2. ✰✰✰✰

3. ✰✰✰✰
✰✰✰✰

_____ _____ _____

Write the quotient. Skip count or use counters to help.

4. $36 \div 4 = $ _____

5. $16 \div 4 = $ _____

6. $28 \div 4 = $ _____

7. $20 \div 4 = $ _____

8. $32 \div 4 = $ _____

9. $24 \div 4 = $ _____

10. $4\overline{)20}$

11. $4\overline{)28}$

12. $4\overline{)16}$

13. $4\overline{)32}$

14. $4\overline{)24}$

15. $4\overline{)4}$

16. $4\overline{)12}$

17. $4\overline{)0}$

18. $4\overline{)8}$

19. $4\overline{)36}$

PROBLEM SOLVING

20. Sonia has a rock collection. She took
24 rocks and put 4 rocks in each box.
How many boxes did she use? _____

21. Peter made 4 boxes to hold all of his 16 model
cars. He put the same number of cars into
each box. How many were in each box? _____

Use with Lesson 5–6, text pages 174–175.

Dividing By 5

Name_____

Date _____

$25 \div 5 = \underline{\ ?\ }$ or $5)\overline{25}^{?}$

Think: $\underline{\ ?\ } \times 5 = 25 \longrightarrow 5 \times 5 = 25$

So $25 \div 5 = \mathbf{5}$, or $5)\overline{25}^{5}$.

Write a division sentence for each model.

1.

2. ■ ■ ■ ■ ■

3.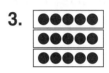

_____

_____

_____

Divide. Skip count or use counters to help.

4. $25 \div 5 = $ ___

5. $10 \div 5 = $ ___

6. $40 \div 5 = $ ___

7. $45 \div 5 = $ ___

8. $20 \div 5 = $ ___

9. $35 \div 5 = $ ___

10. $30 \div 5 = $ ___

11. $5 \div 5 = $ ___

12. $0 \div 5 = $ ___

13. $40 \div 5 = $ ___

14. $15 \div 5 = $ ___

15. $5)\overline{40}$

16. $5)\overline{30}$

17. $5)\overline{5}$

18. $5)\overline{15}$

19. $5)\overline{45}$

20. $5)\overline{25}$

21. $5)\overline{0}$

22. $5)\overline{20}$

23. $5)\overline{35}$

24. $5)\overline{10}$

PROBLEM SOLVING

25. Bernard helped his teacher give out 25 sheets of paper to 5 children. He gave the same number of sheets to each child. At most, how many sheets of paper did each child receive?

_____

26. Louise had 40 crayons. She put 5 crayons into each bag. How many bags did she use?

_____

Dividing Cents

Name_____

Date _____

$$12¢ \div 3 = 4¢ \quad \text{or} \quad 3\overline{)12¢}^{\,4¢}$$

↑ cents in all ↑ number of groups ↑ cents in each group

Divide. Remember to write the cent sign.

1. $4¢ \div 2 =$ _____

2. $8¢ \div 1 =$ _____

3. $18¢ \div 3 =$ _____

4. $10¢ \div 2 =$ _____

5. $32¢ \div 4 =$ _____

6. $15¢ \div 5 =$ _____

7. $5\overline{)3\,5¢}$

8. $3\overline{)9¢}$

9. $1\overline{)3¢}$

10. $4\overline{)2\,4¢}$

11. $2\overline{)1\,4¢}$

12. $3\overline{)2\,4¢}$

13. $5\overline{)1\,0¢}$

14. $4\overline{)1\,6¢}$

Multiply or divide.

15. $7 \times 4¢ =$ _____

16. $9 \times 2¢ =$ _____

17. $9 \times 5¢ =$ _____

18. $9 \times 1¢ =$ _____

19. $4 \times 5¢ =$ _____

20. $9 \times 3¢ =$ _____

21. $21¢ \div 3 =$ _____

22. $18¢ \div 2 =$ _____

23. $30¢ \div 5 =$ _____

PROBLEM SOLVING

24. Donna owed 40¢ for 5 overdue library books. She owed the same amount for each book. How much did she owe per book?

_____

25. Emilio spent 36¢ on 4 used paperback books. Each book cost the same amount. How much did each book cost?

_____

Problem-Solving Strategy: Write a Number Sentence

Name_____

Date_____

Add	• Join sets or quantities.	
Subtract	• Separate, or take away from a set.	• Find part of a set.
	• Compare two sets, or quantities.	• Find how many more are needed.
Multiply	• Join equal sets, or quantities.	
Divide	• Separate a set into equal groups.	• Share a set equally.

Write a number sentence to solve.

1. Germaine learned that humans will normally have 20 "baby" teeth and 32 permanent teeth. How many more permanent teeth than "baby" teeth do humans have?

_____

2. For the fair 12 students used plants in their experiments and 9 students used electricity. How many more students used plants than electricity?

_____

3. On Monday morning, 28 students came to the fair. Then 37 came in the afternoon. How many students came to the fair?

_____

4. Anna used 16 magnets on metal strips. She put groups of 4 magnets on each strip. How many strips did she use?

_____

5. There were 15 children who visited a special exhibit. They visited the exhibit in groups of 3. How many groups were there?

_____

6. The 5 tables at the Bay School science fair each held 4 experiments. How many experiments were there at the fair?

_____

7. Carl used 15 glass items and 17 metal items in his recycling experiment. How many items did he use?

_____

8. Juan planted 8 seeds in each of 2 pots for his experiment. How many seeds did he plant?

_____

Collecting Data

Name_____

Date _____

Which type of art do the most students prefer?

Remember: I = 1 student
          ⊞ = 5 students

Most of the students prefer drawing.

Art	Tally	Total
Painting	⊞ II	7
Sculpture	⊞	5
Drawing	⊞ ⊞ I	11

Use the tally chart above to answer questions 1–2.

1. How many more students prefer painting to sculpture?

_____

2. Which type of art do the fewest students favor?

_____

Complete the tally chart. Then use it to answer questions 3–4.

3. How many people responded to the survey "What is your favorite breakfast food?"

_____

4. Which foods are favored by fewer than 5 people?

_____

Favorite Breakfast	Tally	Total
Hot Oatmeal	III	
Pancakes	⊞ III	
Muffins	⊞ I	
Granola	IIII	
Cold Cereal	⊞ ⊞ I	

Use the list to make a tally chart on a separate sheet of paper. Then answer the questions.

Babs - Summer St.
Keiko - Main St.
Angela - River Rd.
Larry - River Rd.

Margaret - First Ave.
Rick - Summer St.
Bill - Summer St.
Leslie - First Ave.

Dana - First Ave.
Norris - First Ave.
Rena - Main St.
Adam - River Rd.

Ellie - Summer St.
Hal - Summer St.
Cynthia - Summer St.

5. Which street do more than five children live on?

_____

6. How many other children live on the same street Dana lives on?

_____

Making Pictographs

Name _____

Date _____

Complete the tally chart.

1. The tally chart shows the balloons Zina sold at a parade.

Color	Tally	Total
Red	ＩＩＩＩ ＩＩＩＩ	
Blue	ＩＩＩＩ Ｉ	
Purple	ＩＩＩＩ ＩＩＩＩ ＩＩ	
Green	ＩＩＩＩ	
Yellow	ＩＩＩＩ	
Silver	ＩＩＩＩ ＩＩＩ	

Use the tally chart above to make a pictograph.

2.

_____

Red	
Blue	
Purple	
Green	
Yellow	
Silver	

Key: Each _____ = _____ balloons.

PROBLEM SOLVING

3. Of which color did Zina sell the most balloons? the fewest?

4. How many more red balloons than yellow balloons did Zina sell?

_____   _____

Making Bar Graphs

Name_____

Date _____

The tally chart shows one class's favorite animals at the petting zoo.

1. Complete the tally chart.

Animal	Tally	Total
lamb	ЖЖ III	
kid	ЖЖ ЖЖ	
calf	ЖЖ	
colt	ЖЖ ЖЖ II	
rabbit	ЖЖ ЖЖ IIII	

Use the tally chart to make a bar graph.

2. _____

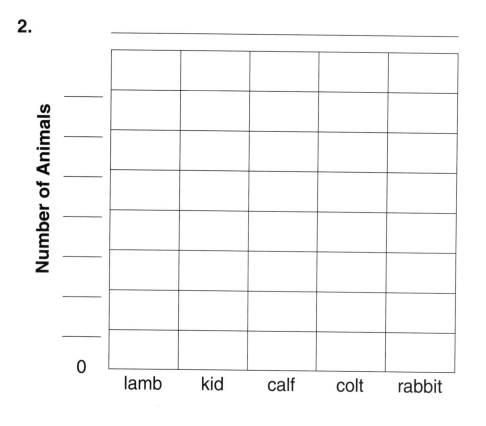

PROBLEM SOLVING

3. Which animal was the favorite of 12 students?

_____

4. Which animal was liked by twice the number of students who liked the calf?

_____

5. Which animal was most popular? _____

Arrangements and Combinations

Name _____

Date _____

John made a list to show how many ways he could make this flag using the colors green, yellow, and red.

Upper	Lower	
red	green	yellow
red	yellow	green
green	yellow	red
green	red	yellow
yellow	red	green
yellow	green	red

PROBLEM SOLVING Make an organized list.

1. Ward is buying a can of juice. He has exactly the right change: 35¢. The machine will not take pennies. How many different combinations of coins could Ward have?

    Q   Q
    D   N
         N

2. Rosie is planting a garden. There are four rows. She wants one row of spinach, two rows of tomatoes, and one row of cucumbers. How many different ways can Rosie plant her garden?

Make an organized list and a tree diagram.

3. Lily has a pair of white shoes and a pair of black shoes to go with her dress. She can wear white socks, lace socks, or blue socks with her shoes. How many ways can Lily wear her socks and shoes?

Probability Experiments

Name _____

Date _____

The chance of landing on red, yellow, or blue is **not** equally likely.

There are 2 blue sections out of a total of 7 sections.

The probability of this spinner landing on blue is 2 out of 7.

The probability of landing on **not** red is 4 out of 7.

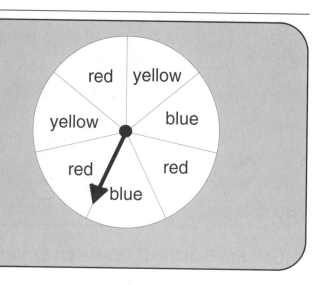

Use the spinner at the right to find the probability of landing on:

1. green _____

2. red _____

3. blue _____

4. yellow _____

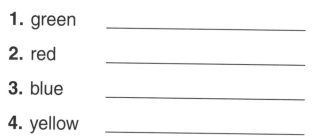

Use the spinner below.

5. What is the probability of the spinner landing on red?

_____

6. What is the probability of the spinner landing on blue?

_____

7. Is it more likely that the spinner will land on blue or green? Explain.

_____

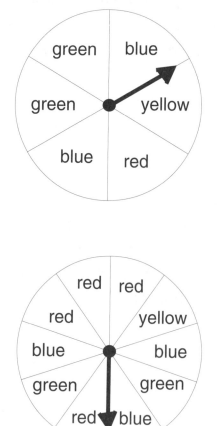

8. What is the probability of the spinner landing on yellow?

_____

9. What is the probability of the spinner landing on green?

_____

Problem-Solving Strategy: Use a Graph

Name_____

Date_____

How many pencils does Ida have in all?

Ida has 12 red pencils, 8 yellow pencils and 16 white pencils in her collection.

Ida has 36 pencils.

Pencils in Ida's Collection	
Red	✎ ✎ ✎
Yellow	✎ ✎
White	✎ ✎ ✎ ✎
Key: Each ✎ = 4 pencils.	

Use the bar graph to answer problems 1–3.

1. On which day was the greatest length of trail cleared?

   _____

2. How many more feet of trail were cleared Thursday than on Tuesday?

   _____

3. On which two days were a total of 350 feet of trail cleared?

   _____

Feet of Trail Cleared

Use the pictograph to answer questions 4–7.

4. On which day were the greatest number of apples sold?

   _____

5. On which day were three times more apples sold than on Friday?

   _____

6. How many symbols would you need to add to increase the number of apples sold on Monday to 48?

Apples Sold	
Mon.	🍎 🍎 🍎 🍎
Tues.	🍎 🍎 🍎 🍎 🍎 🍎
Wed.	🍎 🍎 🍎
Thurs.	🍎 🍎 🍎 🍎 🍎
Fri.	🍎 🍎
Key: Each 🍎 = 6 apples.	

7. What would a half apple represent?

   _____

69

Quarter Inch, Half Inch, Inch

Name _____

Date _____

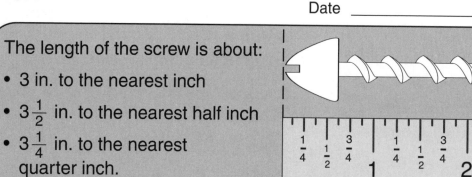

The length of the screw is about:

- 3 in. to the nearest inch
- $3\frac{1}{2}$ in. to the nearest half inch
- $3\frac{1}{4}$ in. to the nearest quarter inch.

Measure each to the nearest inch, half inch, and quarter inch.

1.

_____

2.

_____

Measure the length to the nearest half inch.

3. a book _____

4. an envelope _____

5. a piece of chalk _____

6. a pencil _____

7. a lunch box _____

8. a baseball card _____

Measure the length to the nearest quarter inch.

9. an eraser _____

10. a finger _____

11. a pen _____

12. a shirt cuff _____

Draw a line for each length.

13. 5 in.

14. $4\frac{1}{2}$ in.

15. $2\frac{3}{4}$ in.

Foot, Yard

Name_____

Date_____

1 foot (ft) = 12 inches (in.)	Compare: 10 ft __?__ 3 yd
1 yard (yd) = 36 inches	
1 yard = 3 feet	

ft	3	6	9
yd	1	2	3

9 ft = 3 yd
So 10 ft > 3 yd.

Write the unit used to measure each: in., ft, or yd.

1. the width of a watchband _____

2. the length of a crayon _____

3. the height of a ladder _____

4. the length of a toothbrush _____

5. the length of your arm _____

6. the length of a city block _____

7. the length of the schoolyard _____

8. the height of your kitchen _____

9. the length of a hallway _____

10. the length of a finger _____

Write the letter of the best estimate.

11. the width of a picture window _____   **a.** 6 ft   **b.** 6 yd   **c.** 6 in.

12. the width of your math book _____   **a.** 8 ft   **b.** 8 in.   **c.** 8 yd

13. the length of your finger _____   **a.** 2 yd   **b.** 2 ft   **c.** 2 in.

14. the height of a grandfather clock _____   **a.** 6 ft   **b.** 6 yd   **c.** 6 in.

15. the height of a bench _____   **a.** 2 in.   **b.** 2 ft   **c.** 2 yd

Compare. Write <, =, or >.

16. 5 yd ___ 5 ft   **17.** 7 in. ___ 3 ft   **18.** 3 ft ___ 50 in.

19. 21 ft ___ 4 yd   **20.** 9 yd ___ 9 in.   **21.** 36 in. ___ 3 ft

Mile

Name _____

Date _____

A mile (mi) is about how far you can walk in 25 minutes.

1 mi = 5280 ft
1 mi = 1760 yd

Which unit would you use to measure each: in., ft, yd, or mi?

1. length of Canada – U.S. border

   _____

2. length of a city block

   _____

3. distance from your desk to the classroom door

   _____

4. length of your foot

   _____

Write the letter of the best estimate.

5. length of a driveway _____   **a.** 45 mi   **b.** 45 ft   **c.** 4 mi

6. length of a runway _____   **a.** 4 mi   **b.** 30 mi   **c.** 300 mi

Compare. Write $<, =, >$.

7. 1 mi ____ 1800 yd

8. 6000 ft ____ 2 mi

9. 5280 ft ____ 1 mi

PROBLEM SOLVING Use the map.

10. What is the shortest distance between Freemont and Brookville?

    _____

11. About how far is it to Derby from Ansonia if you go through Charlotte on the way?

    _____

Measurement Sense

Name _____

Date _____

You can estimate to find about how long something is, then measure to check.

The chain of paper clips is about 4 inches.

$\frac{1}{4}$ $\frac{1}{2}$ $\frac{3}{4}$ 1 $\frac{1}{4}$ $\frac{1}{2}$ $\frac{3}{4}$ 2 $\frac{1}{4}$ $\frac{1}{2}$ $\frac{3}{4}$ 3 $\frac{1}{4}$ $\frac{1}{2}$ $\frac{3}{4}$ 4

Match the object to the measurement that describes it.

1. length of an earthworm _____

2. height of a house _____

3. width of a car _____

4. length of a football field _____

a. 30 ft

b. 100 yd

c. 8 in.

d. 6 ft

Estimate, then find the actual measure.
Use a ruler, a yardstick, or a tape measure.

	Estimate	Actual
5. width of this page	_____	_____
6. length of the classroom	_____	_____
7. height of a door	_____	_____

Name two objects having each length.

8. about 8 in.

_____

_____

9. about 8 ft

_____

_____

 73

Customary Units of Capacity

Name_____

Date_____

| 1 cup (c) | 1 pint (pt)
1 pt = 2 c | 1 quart (qt)
1 qt = 2 pt | 1 half gallon
1 half gal = 2 qt | 1 gallon (gal)
1 gal = 4 qt |

Which unit is used to measure each: c, pt, qt, or gal?

1. the amount of juice in a large pitcher   _____

2. the amount of water in a swimming pool   _____

3. the amount of water in a pail   _____

4. the amount of milk in a drinking glass   _____

5. the amount of soup in a bowl   _____

6. the amount of gas in a car   _____

7. the amount of salad oil in a large bottle   _____

8. the amount of cider in a jug   _____

Compare. Write <, =, or >.

9. 4 qt _____ 1 half gallon

10. 8 pt _____ 8 qt

11. 3 gal _____ 30 pt

12. 2 qt _____ 1 half gallon

13. 20 c _____ 25 pt

14. 6 pt _____ 20 c

Complete. You may make a table.

15. 1 qt = _____ pt

16. 3 qt = _____ pt

17. 6 qt = _____ pt

18. 2 pt = _____ qt

19. 8 pt = _____ qt

20. 5 pt = _____ c

21. 1 qt = _____ c

22. 5 qt = _____ c

23. 10 pt = _____ qt

Ounce, Pound

16 oz = 1 lb

About 1 ounce (oz)   About 1 pound (lb)

Which unit is used to measure each: oz or lb?

1. box of cereal _____

2. motorcycle _____

3. dictionary _____

4. telephone _____

5. baseball _____

6. baseball hat _____

Write the letter of the best estimate.

7. canary _____   **a.** 15 oz   **b.** 5 lb   **c.** 5 oz

8. ballerina _____   **a.** 110 lb   **b.** 11 oz   **c.** 11 lb

9. watermelon _____   **a.** 18 oz   **b.** 8 lb   **c.** 1 lb

Compare. Write <, =, or >.

10. 8 oz _____ 1 lb

11. 32 oz _____ 2 lb

12. 64 oz _____ 5 lb

13. 1 lb _____ 12 oz

14. 80 oz _____ 5 lb

15. 112 oz _____ 7 lb

16. 13 oz _____ 1 lb

17. 18 oz _____ 1 lb

18. 3 lb _____ 52 oz

19. 31 lb _____ 30 oz

20. 14 oz _____ 1 lb

21. 16 oz _____ 1 lb

PROBLEM SOLVING

22. Mr. Parker made 4 pounds of potato salad for a party. How many ounces of potato salad did Mr. Parker make?

_____

 75

Metric Units of Length

Name_____

Date _____

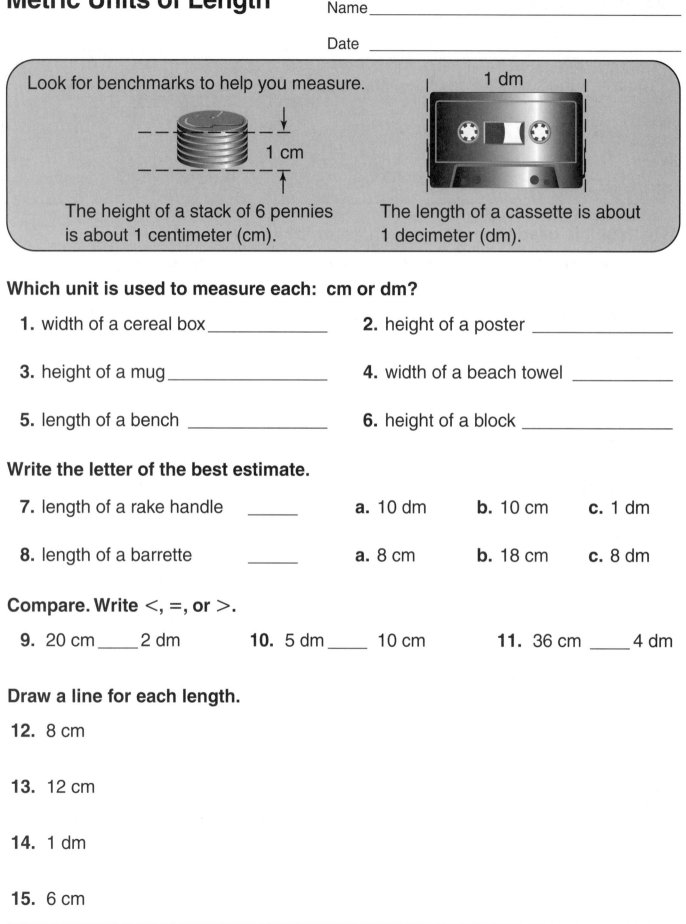

Look for benchmarks to help you measure.

The height of a stack of 6 pennies is about 1 centimeter (cm).

1 dm

The length of a cassette is about 1 decimeter (dm).

Which unit is used to measure each: cm or dm?

1. width of a cereal box _____

2. height of a poster _____

3. height of a mug_____

4. width of a beach towel _____

5. length of a bench _____

6. height of a block _____

Write the letter of the best estimate.

7. length of a rake handle _____   **a.** 10 dm   **b.** 10 cm   **c.** 1 dm

8. length of a barrette _____   **a.** 8 cm   **b.** 18 cm   **c.** 8 dm

Compare. Write <, =, or >.

9. 20 cm_____2 dm   **10.** 5 dm _____ 10 cm   **11.** 36 cm _____4 dm

Draw a line for each length.

12. 8 cm

13. 12 cm

14. 1 dm

15. 6 cm

Meter

Name_____

Date _____

Most doors in your school are probably about 1 meter (m) wide.

1 m = 100 cm
1 m = 10 dm

Which unit is used to measure each: cm, dm, or m?

1. length of a windshield wiper _____

2. height of a schoolbus _____

3. width of a pie pan _____

4. length of a safety pin_____

5. length of a volleyball net _____

6. height of a teacup _____

7. length of a crayon _____

8. length of a boat_____

Write the letter of the best estimate.

9. the height of a stack of 10 dimes _____   **a.** 1 cm   **b.** 1 m   **c.** 1 dm

10. the thickness of a sandwich   _____   **a.** 3 m   **b.** 3 cm   **c.** 15 cm

11. the height of a toy rocket ship   _____   **a.** 4 dm   **b.** 4 m   **c.** 100 m

Compare. Write <, =, or >.

12. 400 cm_____ 4 m

13. 19 dm _____ 200 cm

14. 30 cm ____ 4 dm

15. 55 cm _____ 6 m

16. 2 m _____ 20 dm

17. 88 dm _____8 m

18. 3 m_____ 30 cm

19. 10 dm _____ 1 m

20. 40 dm _____4 m

PROBLEM SOLVING

21. A lilac bush grew to a height of 3 m. When it was planted, the bush was 10 dm tall. How much did the lilac bush grow?

_____

Kilometer

> A kilometer (km) is about how far you can walk in 15 minutes.
>
> | |
> | --- |
> | 1 kilometer = 1000 meters |
> | 1 km = 1000 m |
> | 10 dm = 1 m |
> | 10 cm = 1 dm |

Which unit is used to measure each: cm, dm, m or km?

1. length of a rowboat _____

2. height of a tree _____

3. length of a bread knife _____

4. distance across an ocean _____

5. length of toothbrush _____

6. width of window _____

7. distance from home to school _____

8. length of classroom _____

Compare. Write <, =, or >.

9. 2 km _____ 2000 m

10. 1800 dm _____ 10 m

11. 2700 cm _____ 27 m

PROBLEM SOLVING

12. What is the shortest distance in kilometers from the Base Camp to the Summit Camp?

_____

13. Miranda and Kate started at Base Camp and walked halfway to the Summit Camp for a picnic. How many kilometers did they walk?

_____

14. In the afternoon Kate and Miranda joined Robin in a hike from their picnic spot to Eagle Rock, then on to Summit Camp. How many kilometers did they hike after lunch?

_____

Milliliter, Liter

Name_____

Date_____

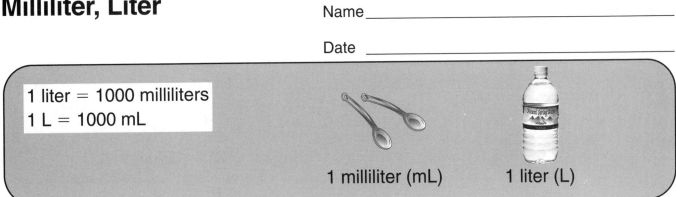

1 liter = 1000 milliliters
1 L = 1000 mL

1 milliliter (mL)   1 liter (L)

**Write *less than a liter* or *more than a liter*
for the amount of liquid each real object holds.**

1. _____

2. _____

3. _____

4. _____

Which unit is used to measure each: mL or L?

5. liquid medicine   _____

6. carton of milk   _____

7. water in a bathtub   _____

8. juice in a baby's bottle   _____

Write the letter of the best estimate.

9. syrup for 2 pancakes   _____   **a.** 25 mL   **b.** 2 mL   **c.** 2 L

10. water in a rain barrel   _____   **a.** 75 mL   **b.** 7 L   **c.** 75 L

Compare. Write <, =, or >.

11. 3640 mL _____ 3 L   **12.** 4 L _____ 8000 mL   **13.** 1 L _____ 1000 mL

Gram, Kilogram

Name _____

Date _____

> 1 kilogram = 1000 grams
> 1kg = 1000 g

1 gram (g)

1 kilogram (kg)

**Write *more than a kilogram* or *less than a kilogram*
for the mass of each real object.**

1. _____

2. _____

3. _____

4. _____

Which unit is used to measure each: g or kg?

5. paper clip _____   **6.** tent _____   **7.** butter for toast _____

Write the letter of the best estimate.

8. medium-sized adult dog	_____	**a.** 14 g	**b.** 14 kg	**c.** 140 kg
9. small paper bag	_____	**a.** 20 g	**b.** 20 kg	**c.** 2 kg
10. car	_____	**a.** 1000 kg	**b.** 100 kg	**c.** 100 g
11. math book	_____	**a.** 50 g	**b.** 1 kg	**c.** 5 kg

Compare. Write <, =, or >.

12. 4000 g _____ 40 kg   **13.** 2 kg _____ 2000 g   **14.** 3300 g _____ 3 kg

Measuring Tools

Name _____

Date _____

Measuring Tools	
meterstick	yardstick
ruler (in.)	ruler (cm)
balance	scale
cup	liter
gallon	tape measure

Write the tool you would use to measure each.

1. mass of a pumpkin

_____

2. height of a mouse

_____

3. amount of oil in a tank truck

_____

4. width of a bed

_____

Match each object with the tool you would use to find each measure.

5. weight of an apple _____

a. meterstick

6. height of a picture frame _____

b. gallon

7. length of a porch _____

c. inch ruler

8. water in a park fountain _____

d. scale

9. Name two things you can measure with each:

meterstick or yardstick _____

_____

inch ruler or centimeter ruler _____

_____

liter or gallon _____

_____

Temperature

Name_____

Date _____

	Celsius (°C)	Fahrenheit (°F)
Normal body temperature	37°C	98.6°F
Water freezes	0°C	32°F

Write the letter of the most reasonable temperature.

1. hot cider   _____   **a.** 125°F   **b.** 25°F   **c.** 45°F

2. snowball   _____   **a.** 32°C   **b.** ⁻5°C   **c.** 50°C

3. room in a house   _____   **a.** 100°C   **b.** 1°C   **c.** 20°C

Answer *Yes* or *No*.

4. The bath water is 95°C. Is it just right for a bath?

_____

5. The temperature in the house is 68°C. Are you comfortable?

_____

6. The school nurse takes your temperature. It is 101°F. Do you have a fever?

_____

7. The temperature outside is 10°C. Could it snow today?

_____

Write each temperature.

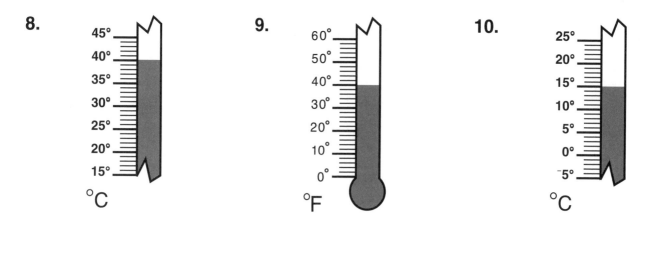

8.

°C

9.

°F

10.

°C

_____   _____   _____

   Use with Lesson 7–13 and 7–14, text pages 234–235.

Quarter Hour

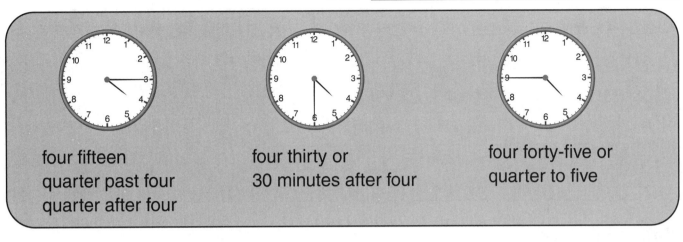

four fifteen
quarter past four
quarter after four

four thirty or
30 minutes after four

four forty-five or
quarter to five

Write each time in standard form.

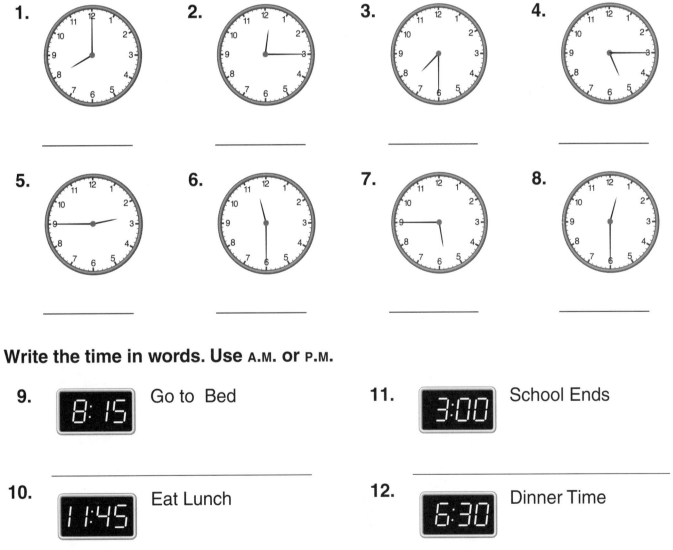

1.

_____

2.

_____

3.

_____

4.

_____

5.

_____

6.

_____

7.

_____

8.

_____

Write the time in words. Use A.M. or P.M.

9. `8:15` Go to Bed

_____

10. `11:45` Eat Lunch

_____

11. `3:00` School Ends

_____

12. `6:30` Dinner Time

_____

Minutes

Name_____

Date _____

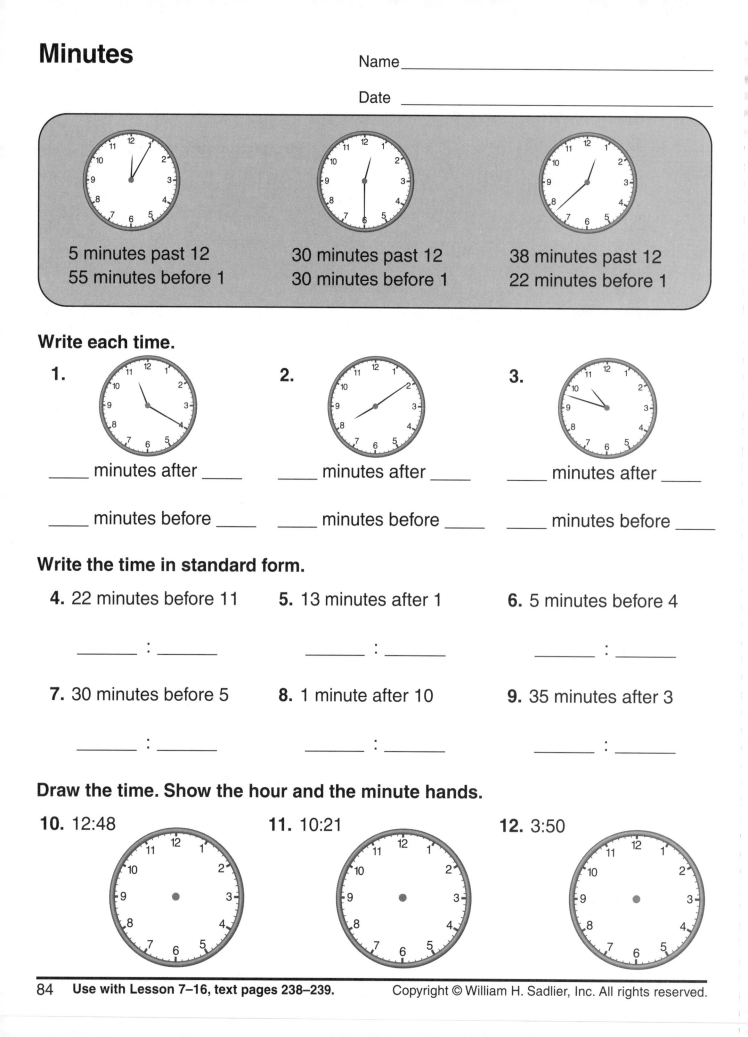

5 minutes past 12
55 minutes before 1

30 minutes past 12
30 minutes before 1

38 minutes past 12
22 minutes before 1

Write each time.

1.

____ minutes after ____

____ minutes before ____

2.

____ minutes after ____

____ minutes before ____

3.

____ minutes after ____

____ minutes before ____

Write the time in standard form.

4. 22 minutes before 11

_____ : _____

5. 13 minutes after 1

_____ : _____

6. 5 minutes before 4

_____ : _____

7. 30 minutes before 5

_____ : _____

8. 1 minute after 10

_____ : _____

9. 35 minutes after 3

_____ : _____

Draw the time. Show the hour and the minute hands.

10. 12:48

11. 10:21

12. 3:50

Elapsed Time

Name _____

Date _____

How much time does the trip take?

| Bus Departs | 7:10 |
| Bus Arrives | 9:30 |

You can skip count minutes by 5 or 10. You can count hours by 1s first.

7:10 }
8:10 } 1 hour
9:10 } 2 hours
9:20 } 10 minutes
9:30 } 20 minutes

This trip takes 2 hours, 20 minutes.

What time will it be in 20 minutes?

1. 2:10 _____   **2.** 8:15 _____   **3.** 7:35 _____

What time will it be in 3 hours?

4. 5:55 _____   **5.** 1:05 _____   **6.** 10:40 _____

What time will it be in 2 hours, 15 minutes?

7. 6:30 _____   **8.** 3:35 _____   **9.** 4:25 _____

How much time is between:

10. 11:15 A.M.   **11.** 2:40 P.M.   **12.** 11:25 P.M.
    12:30 P.M.   8:40 P.M.   2:00 A.M.

_____   _____   _____

PROBLEM SOLVING Use the schedule.

13. How long does it take to go from Lakeland to Grove City?

_____

14. Between what two places will the passengers be at 2:30 P.M.?

_____

Schedule of Bus Stops	
Bolton	12:25 P.M.
Rockdale	1:10 P.M.
N. Orange	2:40 P.M.
S. Orange	2:55 P.M.
Lakeland	3:20 P.M.
Libertad	3:45 P.M.
Grove City	4:45 P.M.

Calendar

Name_____

Date _____

		April				
Sun.	Mon.	Tue.	Wed.	Thurs.	Fri.	Sat.
			1	2	3	4
5	6	7	8	9	10	11
12	13	14	15	16	17	18
19	20	21	22	23	24	25
26	27	28	29	30		

1 week (wk) = 7 days (d)
1 year (yr) = 365 days
      52 weeks
      12 months (mo)
1 leap year = 366 days

Use the calendar above for exercises 1–4.

1. April has _____ days.

2. A leap year has _____ days.

3. Give the date for:

    a. the second Saturday _____

    b. the second day _____

    c. the last Monday _____

    d. the fifth Thursday _____

    e. the second Tuesday _____

    f. the first Sunday of the month _____

4. Give the day of the week for:

    a. April 21 _____

    b. April 25 _____

    c. April 2 _____

    d. May 1 _____

    e. 1 week after April 6 _____

    f. next to last day _____

Write each date two ways.

5. tomorrow's date

_____

_____

6. the last day of this month

_____

_____

Problem-Solving Strategy: Missing Information

Name _____

Date _____

> Elise put $.35 in the dryer at the laundromat for each 15 minutes of drying time. She arrived at 12:00. What time did she leave?
>
> Three groups of 15-minutes equal 45 minutes. So Elise left at 12:45.
>
> Think: You cannot tell what time Elise left until you know how many 15-minute dryer cycles she paid for. Make up a number of cycles: 3

**Write the missing information. Then make it up and solve each problem.
Do your work on a separate sheet of paper.**

1. Elise meets Arlene at the movie theater. Tickets cost $4.25. How much money does Elise need to borrow from Arlene?

   _____

2. After the movie the girls walk to Elise's house to get Arlene's money. They walk 8 blocks. How long does it take them?

   _____

3. At Elise's, the two girls jump rope for half an hour. Then they play in Elise's room. How much longer did they spend in Elise's room than they spent jumping rope?

   _____

4. Arlene's brother drives over to pick her up at Elise's. He is four years older than Arlene. How old is he?

   _____

5. On the way home from school, Arlene and her brother stop at the grocery store. They arrive home at 5:30. How long did it take Arlene and her brother to get home from school?

   _____

6. That night, Elise and Arlene talk on the telephone about what they will do the next day after Elise has her piano lesson and Arlene goes to the dentist. What day of the week will it be?

   _____

Problem Solving:
Review of Strategies

Name _____

Date _____

Solve. Do your work on a separate sheet of paper.

1. Maria and Janessa went to the state fair. Maria went on some rides in the morning and 5 rides in the afternoon. If she went on 8 rides in all, how many rides did she go on in the morning?

   _____

2. Janessa went on 4 rides in the morning. She went on twice that number in the afternoon. How many rides did she go on in all?

   _____

3. Maria brought $9 to the fair. She spent some money on food. Then she had $3.50 left. How much did she spend?

   _____

4. Juan spent $10 at the fair. He spent $4 on rides and $3 on food. How much did he have left to spend on games?

   _____

5. An all-day ticket to the fair costs $7.50. Each day Mark saved 3 quarters and 2 dimes. How long did it take him to save enough money for the ticket?

   _____

6. Every fourth person who came through the gate received a balloon. When the gate opened, 25 people were in line. How many received balloons?

   _____

7. Lavonne bought a hot dog. She gave the cashier $2.00. What was her change?

   _____

8. The roller coaster ride can take 62 people at a time. If 104 people are in line, how many will have to wait?

   _____

9. Maria threw 3 hoops in the hoop toss game each time she played. She played 5 times. Each time she missed one shot. How many hoops did she throw?

   _____

10. Maria and Janessa each won 4 prizes. Tyrell and John each won 3 prizes. Which pair won more prizes? how many more?

   _____

Factors and Products

Name _____

Date _____

$$3 \quad \times \quad 4 \quad = \quad 12$$
$$\underset{\textbf{factors}}{\uparrow \qquad \uparrow} \qquad \underset{\textbf{product}}{\uparrow}$$

$$\begin{array}{r} 4 \leftarrow \\ \times 3 \leftarrow \end{array} \quad \textbf{factors}$$
$$12 \leftarrow \quad \textbf{product}$$

Complete each table.

1.

IN	0	2	5	1	7	3	9
OUT							

□ x 3

2.

IN	4	9	1	2	6	0	3
OUT							

□ x 5

3.

IN	8	5	3	0	7	4	2
OUT							

□ x 2

Write the product.

4. 5	**5.** 4	**6.** 4	**7.** 3	**8.** 5	**9.** 4
×3	×2	×5	×2	×2	×3

10. 4	**11.** 4	**12.** 5	**13.** 2	**14.** 2	**15.** 3
×6	×4	×4	×7	×2	×3

16. 0	**17.** 2	**18.** 4	**19.** 2	**20.** 4	**21.** 5
×3	×9	×7	×3	×8	×6

PROBLEM SOLVING

22. Jennifer collects 4 rocks on each of 7 days.
How many rocks does Jennifer collect in all? _____

Use with Lesson 8–1, text page 254.
89

Multiplying Sixes

Name _____

Date _____

5 groups of 6 are 30.

5 sixes = 30

5 × 6 = 30

groups ⌐ in each   ⌐ in all
         group

$$\begin{array}{r} 6 \\ \times 5 \\ \hline 30 \end{array}$$

☆ ☆ ☆ ☆ ☆ ☆
☆ ☆ ☆ ☆ ☆ ☆
☆ ☆ ☆ ☆ ☆ ☆
☆ ☆ ☆ ☆ ☆ ☆
☆ ☆ ☆ ☆ ☆ ☆

Write the product.

1. $7 \times 6 =$ ____

2. $1 \times 6 =$ ____

3. $2 \times 6 =$ ____

4. $9 \times 6 =$ ____

5. $4 \times 6 =$ ____

6. $3 \times 6 =$ ____

7. $8 \times 6 =$ ____

8. $6 \times 6 =$ ____

Multiply.

9. $\begin{array}{r} 6 \\ \times 3 \\ \hline \end{array}$

10. $\begin{array}{r} 6 \\ \times 7 \\ \hline \end{array}$

11. $\begin{array}{r} 6 \\ \times 8 \\ \hline \end{array}$

12. $\begin{array}{r} 6 \\ \times 2 \\ \hline \end{array}$

13. $\begin{array}{r} 6 \\ \times 1 \\ \hline \end{array}$

14. $\begin{array}{r} 6 \\ \times 5 \\ \hline \end{array}$

15. $\begin{array}{r} 6 \\ \times 4 \\ \hline \end{array}$

16. $\begin{array}{r} 5 \\ \times 6 \\ \hline \end{array}$

17. $\begin{array}{r} 2 \\ \times 6 \\ \hline \end{array}$

18. $\begin{array}{r} 4 \\ \times 6 \\ \hline \end{array}$

19. $\begin{array}{r} 6 \\ \times 6 \\ \hline \end{array}$

20. $\begin{array}{r} 6 \\ \times 9 \\ \hline \end{array}$

21. $\begin{array}{r} 1 \\ \times 6 \\ \hline \end{array}$

22. $\begin{array}{r} 6 \\ \times 0 \\ \hline \end{array}$

23. $\begin{array}{r} 6 \\ \times 7 \\ \hline \end{array}$

24. $\begin{array}{r} 6 \\ \times 6 \\ \hline \end{array}$

25. $\begin{array}{r} 0 \\ \times 6 \\ \hline \end{array}$

26. $\begin{array}{r} 3 \\ \times 6 \\ \hline \end{array}$

PROBLEM SOLVING

27. There are 6 party horns in a pack. How many are there in 3 packs?

_____

28. Maria bought a package of stickers. There were 9 pages with 6 stickers on each page. How many stickers did she buy?

_____

29. There are 6 fish in each of 5 tanks in the pet store. How many fish are there in all?

_____

Multiplying Sevens

Name_____

Date_____

7 groups of 7 are 49.	7		★ ★ ★ ★ ★ ★ ★
	×7		★ ★ ★ ★ ★ ★ ★
7 sevens = 49	49		★ ★ ★ ★ ★ ★ ★
			★ ★ ★ ★ ★ ★ ★
7 × 7 = 49			★ ★ ★ ★ ★ ★ ★
			★ ★ ★ ★ ★ ★ ★
			★ ★ ★ ★ ★ ★ ★

Complete each multiplication fact.

1.

____ × 7 = ____

2.

____ × 7 = ____

3.

____ × 7 = ____

Write the product.

4. $7 \times 7 =$ ____   **5.** $1 \times 7 =$ ____   **6.** $2 \times 7 =$ ____   **7.** $5 \times 7 =$ ____

8. $0 \times 7 =$ ____   **9.** $8 \times 7 =$ ____   **10.** $3 \times 7 =$ ____   **11.** $6 \times 7 =$ ____

12. $9 \times 7 =$ ____   **13.** $7 \times 4 =$ ____   **14.** $7 \times 2 =$ ____   **15.** $4 \times 7 =$ ____

Multiply.

16. 7	**17.** 7	**18.** 7	**19.** 7	**20.** 7	**21.** 7
×3	×7	×8	×2	×1	×5

22. 7	**23.** 7	**24.** 2	**25.** 7	**26.** 7	**27.** 3
×4	×0	×7	×9	×6	×7

PROBLEM SOLVING

28. Gina's swim team took part in 4 events. They won 7 ribbons in each event. How many ribbons did they win in all?

_____

Multiplying Eights

Name _____

Date _____

6 groups of 8 are 48.	8
6 eights = 48	×6
	48
6 × 8 = 48	

Complete each multiplication fact.

1.

_____ × 8 = _____

2.

4 × _____ = 32

3.

_____ × _____ = _____

Write the product.

4. 3 × 8 = _____   **5.** 7 × 8 = _____   **6.** 6 × 8 = _____   **7.** 8 × 8 = _____

8. 5 × 8 = _____   **9.** 9 × 8 = _____   **10.** 2 × 8 = _____   **11.** 1 × 8 = _____

Multiply.

12. 8 ×0   **13.** 8 ×9   **14.** 8 ×8   **15.** 8 ×7   **16.** 8 ×3   **17.** 8 ×1

18. 8 ×4   **19.** 8 ×2   **20.** 8 ×6   **21.** 7 ×8   **22.** 8 ×5   **23.** 3 ×8

PROBLEM SOLVING

24. Carolyn made 8 puppets on Sunday. During the rest of the week she made 3 times as many puppets. How many puppets did Carolyn make during the rest of the week?

_____

Multiplying Nines

Name_____

Date_____

> 2 groups of 9 are 18.
>
> 2 nines = 18
>
> 2 × 9 = 18
>
> $$\begin{array}{r} 9 \\ \times 2 \\ \hline 18 \end{array}$$

Write each multiplication fact.

1. ____ × 9 = ____

2. 3 × ____ = ____

3. ____ × ____ = ____

Write the product.

4. 0 × 9 = ____   **5.** 4 × 9 = ____   **6.** 9 × 9 = ____   **7.** 3 × 9 = ____

8. 6 × 9 = ____   **9.** 7 × 9 = ____   **10.** 5 × 9 = ____   **11.** 1 × 9 = ____

12. 8 × 9 = ____   **13.** 2 × 9 = ____   **14.** 9 × 4 = ____   **15.** 9 × 0 = ____

Multiply.

16. $\begin{array}{r} 9 \\ \times 2 \\ \hline \end{array}$ **17.** $\begin{array}{r} 9 \\ \times 4 \\ \hline \end{array}$ **18.** $\begin{array}{r} 9 \\ \times 8 \\ \hline \end{array}$ **19.** $\begin{array}{r} 9 \\ \times 0 \\ \hline \end{array}$ **20.** $\begin{array}{r} 9 \\ \times 7 \\ \hline \end{array}$ **21.** $\begin{array}{r} 9 \\ \times 9 \\ \hline \end{array}$

PROBLEM SOLVING

22. Jeanie picked 6 crates of strawberries. Each crate holds 9 pints. How many pints of strawberries did Jeanie pick? _____

Multiplying Three Numbers

Multiply factors in parentheses first.

$(3 \times 4) \times 2 = \underline{\ ?\ }$ or $3 \times (4 \times 2) = \underline{\ ?\ }$

$12 \times 2 = \underline{\ ?\ }$ $3 \times 8 = \underline{\ ?\ }$

$12 \times 2 = \boxed{24}$ $3 \times 8 = \boxed{24}$

Multipy. Use the grouping shown.

1. $(2 \times 3) \times 1 = \underline{\hspace{1cm}}$

$\underline{\hspace{1cm}} \times 1 = \underline{\hspace{1cm}}$

2. $(3 \times 2) \times 1 = \underline{\hspace{1cm}}$

$\underline{\hspace{1cm}} \times 1 = \underline{\hspace{1cm}}$

3. $(2 \times 4) \times 1 = \underline{\hspace{1cm}}$

$\underline{\hspace{1cm}} \times 1 = \underline{\hspace{1cm}}$

4. $2 \times (3 \times 1) = \underline{\hspace{1cm}}$

$2 \times \underline{\hspace{1cm}} = \underline{\hspace{1cm}}$

5. $2 \times (4 \times 1) = \underline{\hspace{1cm}}$

$2 \times \underline{\hspace{1cm}} = \underline{\hspace{1cm}}$

6. $(1 \times 5) \times 3 = \underline{\hspace{1cm}}$

$\underline{\hspace{1cm}} \times 3 = \underline{\hspace{1cm}}$

7. $(4 \times 2) \times 3 = \underline{\hspace{1cm}}$

$\underline{\hspace{1cm}} \times 3 = \underline{\hspace{1cm}}$

8. $(7 \times 1) \times 6 = \underline{\hspace{1cm}}$

$\underline{\hspace{1cm}} \times 6 = \underline{\hspace{1cm}}$

9. $(2 \times 4) \times 5 = \underline{\hspace{1cm}}$

$\underline{\hspace{1cm}} \times 5 = \underline{\hspace{1cm}}$

Write the product. Use any grouping.

10. $3 \times 5 \times 1 = \underline{\hspace{1cm}}$

11. $2 \times 4 \times 3 = \underline{\hspace{1cm}}$

12. $3 \times 0 \times 9 = \underline{\hspace{1cm}}$

13. $8 \times 1 \times 2 = \underline{\hspace{1cm}}$

14. $3 \times 3 \times 5 = \underline{\hspace{1cm}}$

15. $4 \times 2 \times 3 = \underline{\hspace{1cm}}$

16. $2 \times 4 \times 2 = \underline{\hspace{1cm}}$

17. $6 \times 1 \times 4 = \underline{\hspace{1cm}}$

18. $5 \times 2 \times 4 = \underline{\hspace{1cm}}$

19. $5 \times 2 \times 2 = \underline{\hspace{1cm}}$

20. $8 \times 2 \times 0 = \underline{\hspace{1cm}}$

21. $4 \times 5 \times 1 = \underline{\hspace{1cm}}$

22. $3 \times 3 \times 2 = \underline{\hspace{1cm}}$

23. $1 \times 7 \times 2 = \underline{\hspace{1cm}}$

24. $6 \times 2 \times 2 = \underline{\hspace{1cm}}$

PROBLEM SOLVING

25. There are 3 books on each shelf of the bookcases. Each bookcase has 3 shelves. How many books are there in 4 bookcases?

_____

Division Review

Name_____

Date _____

quotient

$$12 \div 4 = 3 \quad \text{or} \quad 4\overline{)12}^{\;3}$$

divisor

dividend

Divide.

1. $1\overline{)5}$ **2.** $2\overline{)18}$ **3.** $3\overline{)21}$ **4.** $3\overline{)12}$ **5.** $5\overline{)20}$

6. $2\overline{)2}$ **7.** $4\overline{)4}$ **8.** $4\overline{)24}$ **9.** $5\overline{)35}$ **10.** $4\overline{)28}$

11. $5\overline{)30}$ **12.** $3\overline{)27}$ **13.** $4\overline{)32}$ **14.** $5\overline{)25}$ **15.** $4\overline{)36}$

16. $4\overline{)20}$ **17.** $5\overline{)45}$ **18.** $3\overline{)24}$ **19.** $4\overline{)16}$ **20.** $5\overline{)40}$

Complete.

21. $21 \div 3 = $ _____

$7 \times 3 = $ _____

22. $18 \div 3 = $ _____

$3 \times $ _____ $= 18$

23. $30 \div 5 = $ _____

_____ $\times 5 = 30$

24. $32 \div $ _____ $= 8$

_____ $\times 4 = 32$

25. $16 \div $ _____ $= 4$

$4 \times $ _____ $= 16$

26. $4 \div 1 = $ _____

_____ $\times 4 = 4$

27. $12 \div 4 = $ _____

$3 \times 4 = $ _____

28. $10 \div 2 = $ _____

$5 \times 2 = $ _____

29. $36 \div 4 = $ _____

_____ $\times 4 = 36$

30. $27 \div 3 = $ _____

_____ $\times 3 = 27$

31. $35 \div 5 = $ _____

_____ $\times 5 = 35$

32. $45 \div $ _____ $= 9$

$9 \times $ _____ $= 45$

PROBLEM SOLVING

33. Vida's bookcase holds 45 books. If she places
her books equally on each of the 5 shelves,
how many books are on each shelf?

_____

Dividing by 6

$30 \div 6 = \underline{?}$

or

$\begin{array}{r} ? \\ 6\overline{)30} \end{array}$

Think: $30 \div 6 = \underline{?}$

$\underline{?} \times 6 = 30$

$5 \times 6 = 30$

So $30 \div 6 = 5$ or $6\overline{)30}^{\,5}$.

Write the division fact to find the number of groups.
Then write the related multiplication fact.

1.

$12 \div \underline{\quad} = \underline{\quad}$

$\underline{\quad} \times \underline{\quad} = \underline{\quad}$

2.

$\underline{\quad} \div \underline{\quad} = \underline{\quad}$

$\underline{\quad} \times \underline{\quad} = \underline{\quad}$

3.

$\underline{\quad} \div \underline{\quad} = \underline{\quad}$

$\underline{\quad} \times \underline{\quad} = \underline{\quad}$

Write the quotient. Use counters or skip count to help.

4. $36 \div 6 = \underline{\quad}$ **5.** $42 \div 6 = \underline{\quad}$ **6.** $30 \div 6 = \underline{\quad}$ **7.** $18 \div 6 = \underline{\quad}$

8. $54 \div 6 = \underline{\quad}$ **9.** $12 \div 6 = \underline{\quad}$ **10.** $48 \div 6 = \underline{\quad}$ **11.** $24 \div 6 = \underline{\quad}$

12. $6\overline{)12}$ **13.** $6\overline{)6}$ **14.** $6\overline{)54}$ **15.** $6\overline{)48}$ **16.** $6\overline{)18}$

17. $6\overline{)24}$ **18.** $6\overline{)0}$ **19.** $6\overline{)30}$ **20.** $5\overline{)30}$ **21.** $6\overline{)42}$

PROBLEM SOLVING

22. Fred collected 48 baseball cards. He had
6 cards of each player. How many players
were pictured in Fred's collection?   _____

Dividing by 7

$63 \div 7 =$ ___?___

or

$$\dfrac{?}{7)63}$$

Think: $63 \div 7 =$ ___?___

___?___ $\times 7 = 63$

$9 \times 7 = 63$

So $63 \div 7 = 9$ or $7\overline{)63}.$ with 9 above

Write the division fact to find the number of groups.
Then write the related multiplication fact.

1.

___ $\div 7 =$ ___

___ $\times$ ___ $=$ ___

2.

___ $\div$ ___ $=$ ___

___ $\times$ ___ $=$ ___

3.

___ $\div$ ___ $=$ ___

___ $\times$ ___ $=$ ___

Write the quotient. Use counters or skip count to help.

4. $35 \div 7 =$ ___

5. $63 \div 7 =$ ___

6. $56 \div 7 =$ ___

7. $42 \div 7 =$ ___

8. $14 \div 7 =$ ___

9. $0 \div 7 =$ ___

10. $49 \div 7 =$ ___

11. $7 \div 7 =$ ___

12. $7\overline{)14}$

13. $7\overline{)56}$

14. $7\overline{)49}$

15. $7\overline{)7}$

16. $7\overline{)21}$

17. $7\overline{)28}$

18. $4\overline{)28}$

19. $7\overline{)42}$

20. $6\overline{)42}$

21. $1\overline{)7}$

PROBLEM SOLVING

22. There are 63 apartments in all. There are 7 apartments on each floor. How many floors are there? _____

23. Seven girls grew 21 tomato plants. If each girl grew the same number of plants, how many tomato plants did each girl grow? _____

Use with Lesson 8–9, text pages 268–269.

Dividing by 8

$72 \div 8 = \underline{?}$

or

$\overset{?}{8\overline{)72}}$

Think: $72 \div 8 = \underline{?}$

$\underline{?} \times 8 = 72$

$9 \times 8 = 72$

So $72 \div 8 = 9$ or $8\overset{9}{\overline{)72}}$.

Write the division fact to find the number of groups.
Then write the related multiplication fact.

1.

2.

3.

$40 \div \underline{\hspace{1cm}} = \underline{\hspace{1cm}}$

$\underline{\hspace{1cm}} \times \underline{\hspace{1cm}} = \underline{\hspace{1cm}}$

$\underline{\hspace{1cm}} \div \underline{\hspace{1cm}} = \underline{\hspace{1cm}}$

$\underline{\hspace{1cm}} \times \underline{\hspace{1cm}} = \underline{\hspace{1cm}}$

$\underline{\hspace{1cm}} \div \underline{\hspace{1cm}} = \underline{\hspace{1cm}}$

$\underline{\hspace{1cm}} \times \underline{\hspace{1cm}} = \underline{\hspace{1cm}}$

Write the quotient. Use counters or skip count to help.

4. $32 \div 8 = \underline{\hspace{1cm}}$

5. $16 \div 8 = \underline{\hspace{1cm}}$

6. $64 \div 8 = \underline{\hspace{1cm}}$

7. $56 \div 8 = \underline{\hspace{1cm}}$

8. $72 \div 8 = \underline{\hspace{1cm}}$

9. $48 \div 8 = \underline{\hspace{1cm}}$

10. $8 \div 8 = \underline{\hspace{1cm}}$

11. $24 \div 8 = \underline{\hspace{1cm}}$

12. $8\overline{)40}$

13. $8\overline{)72}$

14. $8\overline{)64}$

15. $8\overline{)8}$

16. $8\overline{)16}$

17. $8\overline{)24}$

18. $8\overline{)48}$

19. $8\overline{)56}$

20. $8\overline{)32}$

21. $4\overline{)32}$

PROBLEM SOLVING

22. A bag is filled with 24 crackers. There are 8 children. How many crackers can each child have if they share equally?

_____

23. Ted drives a truck 40 hours a week. If he drives 8 hours a day, how many days does he work each week?

_____

Dividing by 9

Name_____

Date_____

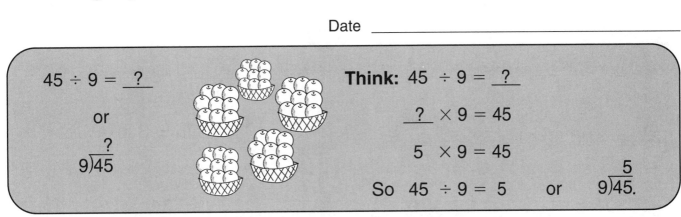

$45 \div 9 = \underline{\ ?\ }$

or

$9\overline{)45}^{\,?}$

Think: $45 \div 9 = \underline{\ ?\ }$

$\underline{\ ?\ } \times 9 = 45$

$5 \times 9 = 45$

So $45 \div 9 = 5$ or $9\overline{)45}^{\,5}$.

**Write the division fact to find the number of groups.
Then write the related multiplication fact.**

1.   2.   3.

$54 \div \underline{\ \ \ \ } = \underline{\ \ \ \ }$ $\underline{\ \ \ \ } \div \underline{\ \ \ \ } = \underline{\ \ \ \ }$ $\underline{\ \ \ \ } \div \underline{\ \ \ \ } = \underline{\ \ \ \ }$

$\underline{\ \ \ \ } \times \underline{\ \ \ \ } = \underline{\ \ \ \ }$ $\underline{\ \ \ \ } \times \underline{\ \ \ \ } = \underline{\ \ \ \ }$ $\underline{\ \ \ \ } \times \underline{\ \ \ \ } = \underline{\ \ \ \ }$

Write the quotient. Use counters or skip count to help.

4. $27 \div 9 = \underline{\ \ \ \ }$ **5.** $63 \div 9 = \underline{\ \ \ \ }$ **6.** $36 \div 9 = \underline{\ \ \ \ }$ **7.** $18 \div 9 = \underline{\ \ \ \ }$

8. $72 \div 9 = \underline{\ \ \ \ }$ **9.** $45 \div 9 = \underline{\ \ \ \ }$ **10.** $0 \div 9 = \underline{\ \ \ \ }$ **11.** $36 \div 9 = \underline{\ \ \ \ }$

12. $9\overline{)72}$ **13.** $9\overline{)27}$ **14.** $9\overline{)36}$ **15.** $9\overline{)81}$ **16.** $9\overline{)9}$

17. $9\overline{)18}$ **18.** $9\overline{)63}$ **19.** $1\overline{)9}$ **20.** $9\overline{)54}$ **21.** $9\overline{)45}$

PROBLEM SOLVING

22. There are 81 packs of light bulbs. If there are
9 packs per case, how many cases are there?   _____

23. There are 18 boys and girls playing a baseball
game. There are 9 players on each team.
How many teams are there?   _____

Operation Patterns

Name _____

Date _____

What number comes next?

18,　　16,　　19,　　17,　　?_____

　　− 2　　　+ 3　　　− 2　　　+ 3

Rule: Subtract 2, Add 3.

18,　　　16,　　　19,　　　17,　　　_20_

Think: 17 + 3 = 20

Write the rule. Then complete the pattern.

1. 25, 22, 19, 16, ____　　　　_____

2. 1, 2, 3, 6, 7, ____　　　　　_____

3. 9, 11, 10, 12, 11, 13, ____　　_____

4. 18, 9, 10, 5, 6, ____　　　　_____

5. 23, 18, 24, 19, 25, ____　　　_____

6. 1, 8, 2, 16, 4, ____　　　　　_____

Complete each pattern.

7.　8　　　8　　　8　　　8
　　　×3　　×4　　×5　　×6

8.　9　　　8　　　7　　　6
　　　×5　　×5　　×5　　×5

9. 6)54　　6)48　　6)42　　**10.** 9)27　　9)36　　9)45

PROBLEM SOLVING

11. April 4 is on a Sunday. What are the dates
of the next two Sundays?　　　　_____

100　**Use with Lesson 8–12, text pages 274–275.**

Fact Families

Name_____

Date _____

These sentences form a fact family for multiplication and division.

$$2 \times 5 = 10 \qquad 10 \div 5 = 2$$

| groups | in each | in all | in all | in each | groups |

$$5 \times 2 = 10 \qquad 10 \div 2 = 5$$

Complete each fact family.

1. $6 \times \rule{2em}{0.4pt} = 54$

$\rule{2em}{0.4pt} \div 9 = 6$

$9 \times \rule{2em}{0.4pt} = 54$

$54 \div 6 = \rule{2em}{0.4pt}$

2. $\rule{2em}{0.4pt} \times 3 = 24$

$24 \div \rule{2em}{0.4pt} = 8$

$3 \times 8 = \rule{2em}{0.4pt}$

$\rule{2em}{0.4pt} \div 8 = 3$

3. $4 \times 6 = \rule{2em}{0.4pt}$

$24 \div \rule{2em}{0.4pt} = 4$

$\rule{2em}{0.4pt} \times 4 = 24$

$\rule{2em}{0.4pt} \div 4 = 6$

4. $7 \times 8 = \rule{2em}{0.4pt}$

$56 \div 8 = \rule{2em}{0.4pt}$

$\rule{2em}{0.4pt} \times 7 = 56$

$56 \div \rule{2em}{0.4pt} = 8$

5. $\rule{2em}{0.4pt} \times 5 = 45$

$\rule{2em}{0.4pt} \div 5 = 9$

$\rule{2em}{0.4pt} \times 9 = 45$

$45 \div \rule{2em}{0.4pt} = 5$

6. $3 \times \rule{2em}{0.4pt} = 21$

$21 \div 7 = \rule{2em}{0.4pt}$

$\rule{2em}{0.4pt} \times 3 = 21$

$21 \div 3 = \rule{2em}{0.4pt}$

Write the complete fact family for each.

7. 2, 9, 18

8. 4, 5, 20

9. 3, 6, 18

PROBLEM SOLVING

10. Tamika collected 72 coins. She put 8 coins in each holder. How many holders did she use? _____

11. Nine students each received 7 stars from the teacher. How many stars did the teacher give to these students? _____

Problem Solving Strategy: Guess and Test

Name_____

Date _____

The sum of two numbers is 9. Their product is 18. What are the two numbers?

Think: _?_ + _?_ = 9; _?_ × _?_ = 18. Guess and test. Use a table.

Sum	Product
1 + 8 = 9	1 × 8 = 8
2 + 7 = 9	2 × 7 = 14
3 + 6 = 9	3 × 6 = 18

The two numbers are 3 and 6.

Solve. Do your work on a separate sheet of paper.

1. Chris bought a card for 60¢. He gave the clerk 6 coins. Which coins did he give the clerk?

   _____

2. The product of two numbers is 36. The sum of the numbers is 13. What are the two numbers?

   _____

3. Reid found 61¢. He found 6 coins. What coins did he find?

   _____

4. The sum of two numbers is 16. The product is 64. Name the numbers.

   _____

5. Max went to the fair with 5 friends. Some paid $2 each to ride the bumper cars. The rest paid $3 each to ride the roller coaster. They spent $16 in all. How many rode each ride?

   _____

6. The quotient of two numbers is 8. The sum of the same two numbers is 72. What are the two numbers?

7. If Angel plants the same number of pepper plants in each of 5 rows, he will have none left over. If he plants 1 in the first row and then 1 more in each row than he did in the previous row, he will plant peppers in only 4 rows. How many plants does he have?

   _____

8. Paul has 30 postcards. He has 4 times as many large postcards as small postcards. How many small postcards does he have? How many large postcards?

   _____

Polygons and Circles

Name_____

Date _____

Polygons   **Circle**

side ←———→   A circle has 0 sides
and 0 corners.

←———corner———→

Write the number of sides and corners for each.

1.
Sides _____
Corners _____

2.
Sides _____
Corners _____

3.
Sides _____
Corners _____

4.
Sides _____
Corners _____

5.
Sides _____
Corners _____

6.
Sides _____
Corners _____

7.
Sides _____
Corners _____

8.
Sides _____
Corners _____

Is each figure a polygon? Write *Yes* or *No*.

9.
_____

10.
_____

11.
_____

12.
_____

Draw a polygon. Answer the question.

13. 4 sides

14. 6 corners

How many corners? _____   How many sides? _____

Lines

Name_____

Date _____

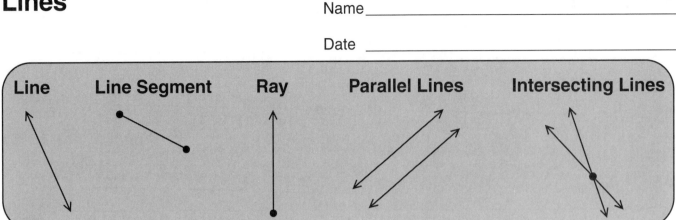

Name each: line, line segment, ray, or none of these.

1.

2.

3.

4.

_____   _____   _____   _____

5.

6.

7.

8.

_____   _____   _____   _____

Draw a line that:

9. is parallel to the line.   **10.** intersects the line.   **11.** intersects both lines.

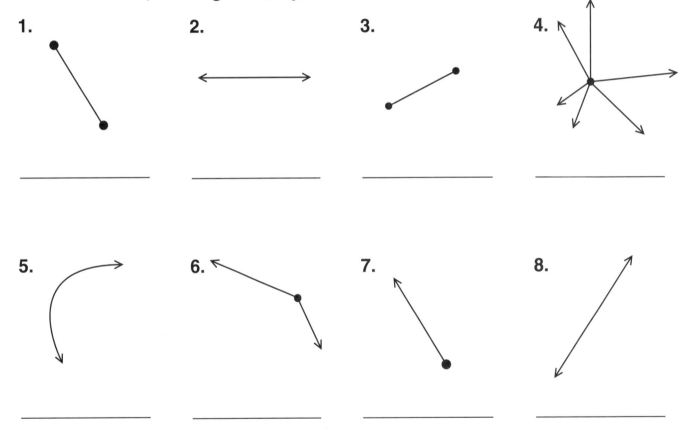

Angles

Name _____

Date _____

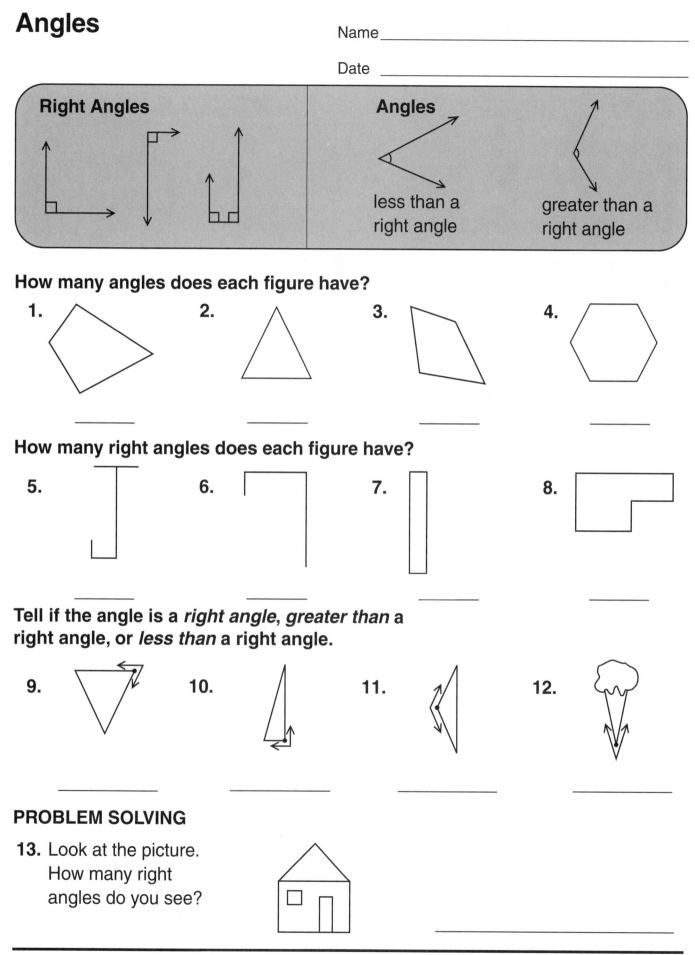

Right Angles

Angles

less than a
right angle

greater than a
right angle

How many angles does each figure have?

1. _____

2. _____

3. _____

4. _____

How many right angles does each figure have?

5. _____

6. _____

7. _____

8. _____

Tell if the angle is a *right angle*, *greater than* a right angle, or *less than* a right angle.

9. _____

10. _____

11. _____

12. _____

PROBLEM SOLVING

13. Look at the picture. How many right angles do you see?

_____

Congruent and Similar Figures

Name_____

Date _____

Congruent Figures	Similar Figures

Do the pairs look congruent? Write *Yes* or *No*.

1.

2.

3.

_____   _____   _____

Draw a congruent figure for each.

4.

5.

6.

7.

Do the pairs look similar? Write *Yes* or *No*.

8.

9.

10.

_____   _____   _____

Draw a similar figure for each.

11.

12.

Use with Lessons 9–4 and 9–5, text pages 296–299.

Ordered Pairs

Name _____

Date _____

Look at the graph below. What picture is at the point (5, 6)?
Find 0. Move 5 spaces to the right. Move 6 spaces up. The
trees are located at (5, 6).

Use the graph. Write the place for each ordered pair.

1. (6, 9) _____

2. (4, 2) _____

3. (5, 5) _____

4. (8, 3) _____

5. (0, 3) _____

6. (9, 5) _____

7. (2, 1) _____

8. (7, 8) _____

9. (2, 5) _____

10. (3, 3) _____

11. (3, 7) _____

12. (5, 0) _____

13. (1, 8) _____

14. (8, 7) _____

15. (5, 4) _____

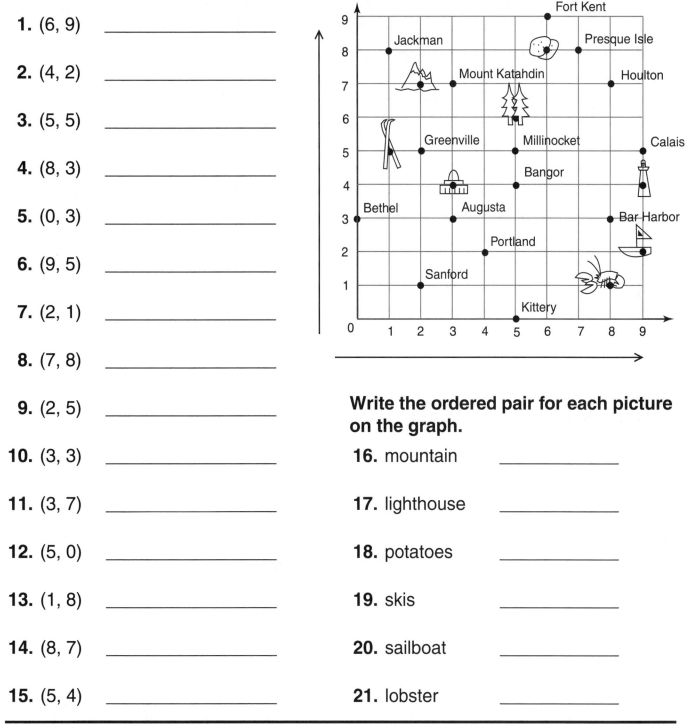

Write the ordered pair for each picture on the graph.

16. mountain _____

17. lighthouse _____

18. potatoes _____

19. skis _____

20. sailboat _____

21. lobster _____

Use with Lesson 9–6, text pages 300–301.

107

Symmetry

Name_____

Date _____

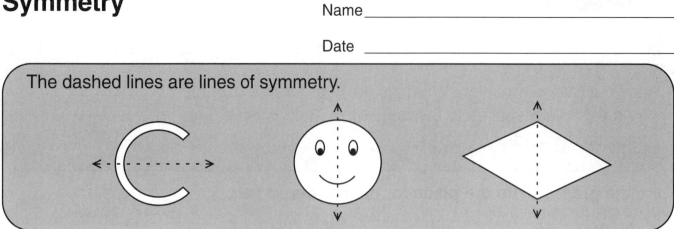

The dashed lines are lines of symmetry.

Is the dashed line a line of symmetry? Write *Yes* or *No*.

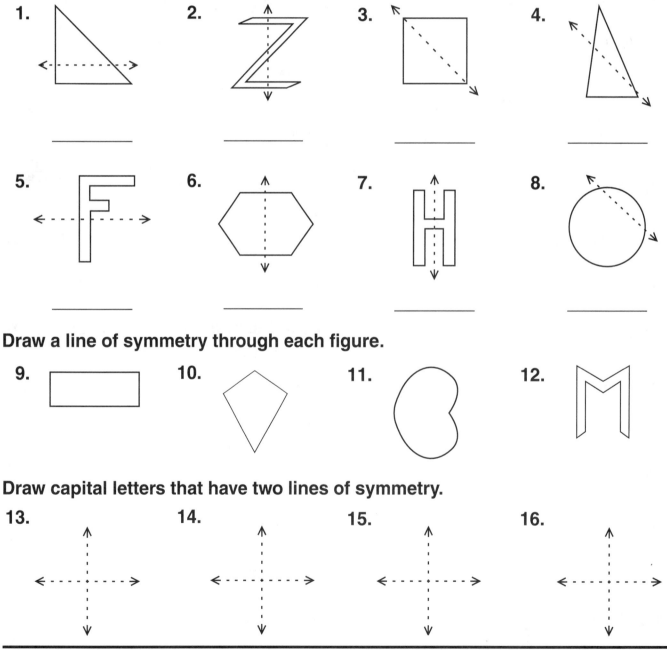

1. _____

2. _____

3. _____

4. _____

5. _____

6. _____

7. _____

8. _____

Draw a line of symmetry through each figure.

9.   10.   11.   12.

Draw capital letters that have two lines of symmetry.

13.   14.   15.   16.

Slides, Flips, and Turns

Name _____

Date _____

Slide	Flip	Turn
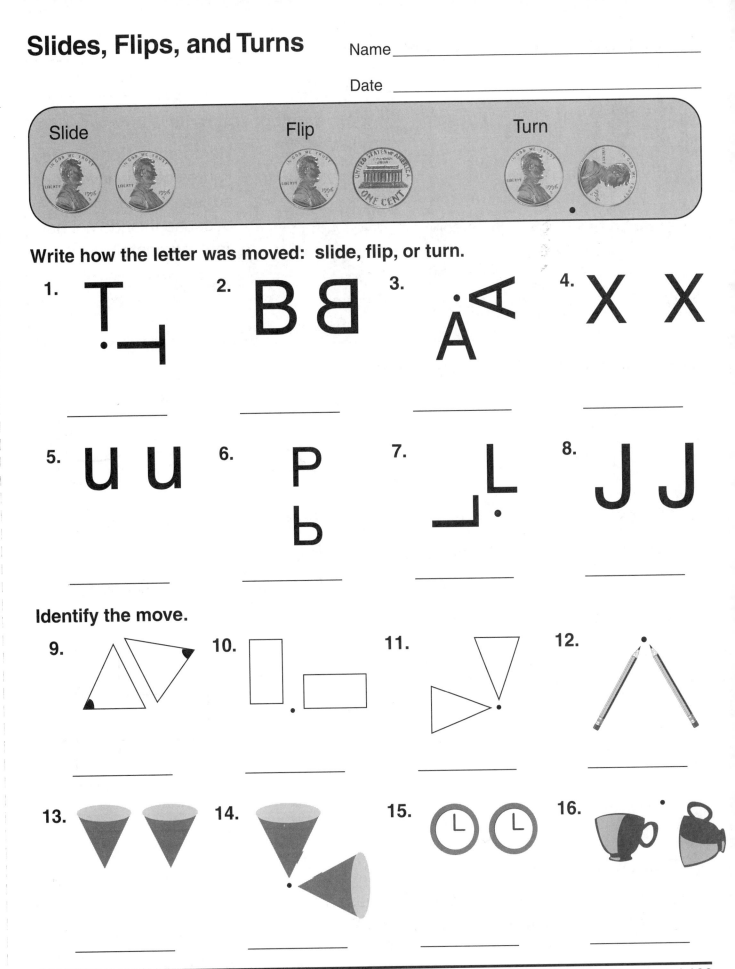		

Write how the letter was moved: slide, flip, or turn.

1. T T
_____

2. B B
_____

3. A A
_____

4. X X
_____

5. u u
_____

6. P b
_____

7. L L
_____

8. J J
_____

Identify the move.

9. _____

10. _____

11. _____

12. _____

13. _____

14. _____

15. _____

16. _____

Exploring Space Figures

Name _____

Date _____

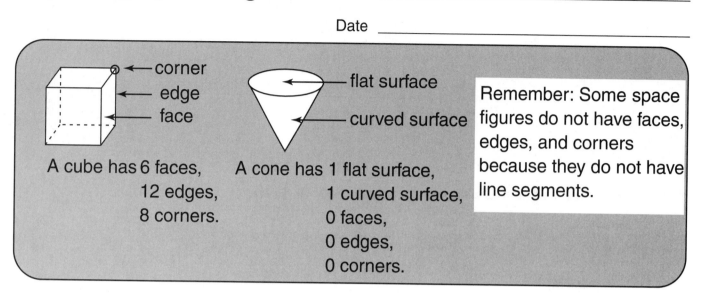

corner
edge
face

A cube has 6 faces,
12 edges,
8 corners.

flat surface
curved surface

A cone has 1 flat surface,
1 curved surface,
0 faces,
0 edges,
0 corners.

Remember: Some space figures do not have faces, edges, and corners because they do not have line segments.

Write the number of faces, edges, and corners.

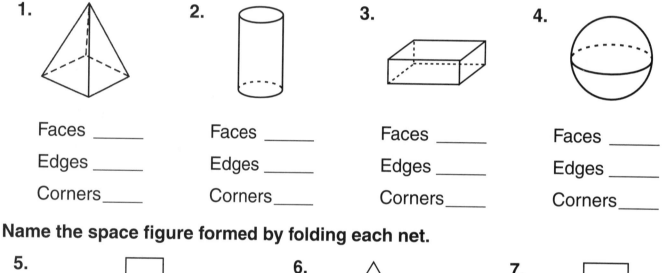

1.

Faces _____

Edges _____

Corners_____

2.

Faces _____

Edges _____

Corners_____

3.

Faces _____

Edges _____

Corners_____

4.

Faces _____

Edges _____

Corners_____

Name the space figure formed by folding each net.

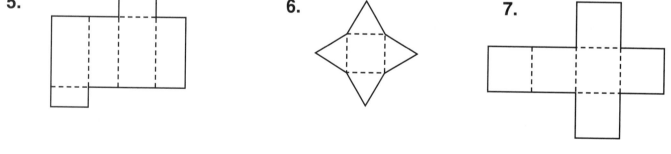

5.

_____

6.

_____

7.

_____

PROBLEM SOLVING

8. I am a space figure with 5 faces, 8 edges, and 4 corners. Which space figure am I?

_____

Exploring Perimeter

Name_____

Date _____

Find the perimeter.

25 ft

10 ft _____ 10 ft

25 ft

To find the perimeter of a figure, add the lengths of its sides.

Perimeter = 25 ft + 10 ft + 25 ft + 10 ft
Perimeter = 70 ft

Write the perimeter.

1. 3 m / 5 m / 4 m / 6 m

_____

2. 5 ft / 5 ft / 5 ft / 5 ft

_____

3. 4 m / 4 m / 4 m / 4 m / 6 m

_____

4. 8 yd / 4 yd / 4 yd / 8 yd

_____

5. 5 in. / 5 in. / 4 in.

_____

6. 3 m / 6 m / 3 m / 3 m / 6 m / 3 m

_____

Use your centimeter ruler to find the perimeter of each.

7.

_____

8.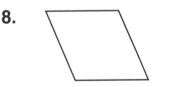

_____

9.

_____

PROBLEM SOLVING

10. Matt has a square picture that measures 10 cm on a side. What is the perimeter of the picture?

_____

11. What is the perimeter of a clock in the shape of a hexagon if each side measures 9 in.?

_____

Use with Lesson 9–10, text pages 306–307.

111

Area

Name _____

Date _____

Area is the number of square units
needed to cover a surface.

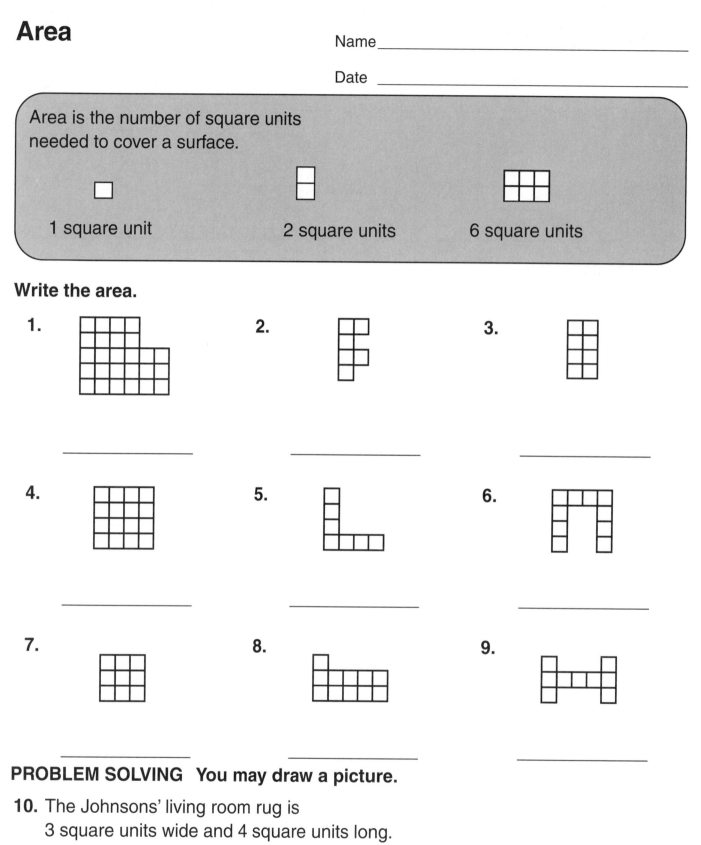

□   1 square unit

2 square units

6 square units

Write the area.

1. _____

2. _____

3. _____

4. _____

5. _____

6. _____

7. _____

8. _____

9. _____

PROBLEM SOLVING   **You may draw a picture.**

10. The Johnsons' living room rug is
3 square units wide and 4 square units long.
What is the area of the rug? _____

11. Bill's bedroom floor is 9 square units long
and 8 square units wide. What is its area? _____

Volume

Name _____

Date _____

The volume of a space figure is the number of cubic units it contains.

1 cubic unit

Volume = 4 cubic units

Volume = 6 cubic units

Write the volume.

1.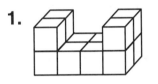

_____

2.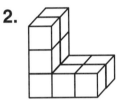

_____

3.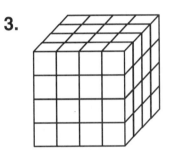

_____

4.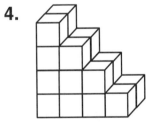

_____

5.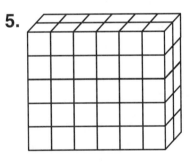

_____

6.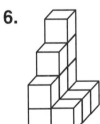

_____

PROBLEM SOLVING

7. How many cubic units are needed to make the figure in exercise 1 a rectangular prism? _____

Problem-Solving Strategy: Logical Reasoning

Name _____

Date _____

Grace's birthday is between May 5 and May 12. It is before the 11th but after the 8th and it is on an even-numbered day. When is Grace's birthday?

Think: Make a chart to show the week and mark off the days it cannot be.

MAY

5̶	6̶	7̶	8̶	9	10	11	1̶2̶

Grace's birthday is between the 8th and the 12th. It is not on an odd-numbered day, so it is not May 9th or May 11th. Grace's birthday must be on May 10.

Solve. Do your work on a separate sheet of paper.

1. Winnie, Delores, Alonzo and Zeke are seated in a row in class. Neither Winnie nor Alonzo is next to the window. Zeke sits between the two girls. In what order are the four seated from the window?

2. Arlene begins school during the week that starts on Sunday, September 15. The date is an even-number whose digits, added together, make a sum that equals half the number. What day does school begin?

3. Tory, Heather, Kara, and Donna went to see a movie. Heather took the first seat. Kara and Tory did not sit next to each other. Donna only sat next to Tory. In what order did the girls sit?

4. Jamal, Liu, Mary, and Tasha are standing in a line. Liu is last. Tasha isn't first or third. Mary is not first. In what order are they standing?

5. Bill, Ray and Nan all live on Maple Street. Bill and Nan live on the same side of the street at odd-numbered addresses. The digits of Bill's address add up to 6. Ray lives across the street from Bill. The addresses are 302, 303, and 309. Who lives where?

6. Lamar, Tanya, and Maria entered a pie-baking contest. Of their three pies, Lamar did not bake the apple pie. Maria baked a berry pie but did not use blueberries. The third pie was made with blackberries. Who made each pie?

Estimating Products

Name_____

Date _____

Rounding:	**Front–End Estimation:**
Round the greater factor to its greatest place value. Then multiply.	Use the value of the front digit of the greater factor. Then multiply.
$2.76 → $3.00 × 3 × 3 about $9.00	$2.76 → $2.00 × 3 × 3 about $6.00

Multiply mentally. Look for a pattern.

1. $3 \times 200 =$ _____   **2.** $3 \times 300 =$ _____   **3.** $3 \times 400 =$ _____

4. $3 \times 500 =$ _____   **5.** $3 \times 600 =$ _____   **6.** $3 \times 700 =$ _____

Estimate by rounding.

7. 2 3 × 3	**8.** 4 4 × 5	**9.** 3 6 × 2	**10.** $.7 8 × 7	**11.** 1 6¢ × 4	**12.** $.6 6 × 6

13. 4 5 3 × 4	**14.** 2 6 9 × 3	**15.** $6.7 1 × 5	**16.** $3.9 3 × 2	**17.** $5.2 7 × 6	**18.** $1.8 1 × 2

Estimate. Use front-end estimation.

19. 5 1 × 3	**20.** 5 7 × 4	**21.** 7 9 × 5	**22.** 3 3¢ × 2	**23.** $.8 4 × 3	**24.** $.5 8 × 6

25. 4 8 2 × 2	**26.** 6 4 8 × 3	**27.** 3 1 7 × 5	**28.** $ 9.4 3 × 2	**29.** $8.7 1 × 4	**30.** $7.3 6 × 4

PROBLEM SOLVING

31. Brigitta bought 3 pairs of socks. Each pair cost $2.29. **About** how much money did she spend?

_____

32. Justin bought 2 tee-shirts. Each shirt cost $3.69. **About** how much did he spend?

_____

 115

Multiplying Two Digits

Name_____

Date _____

Find the product.

1. 4 1
 $\times$ 2

2. 1 2
 $\times$ 3

3. 2 3
 $\times$ 3

4. 3 3
 $\times$ 2

5. 4 2
 $\times$ 2

6. 1 1
 $\times$ 6

7. 2 0
 $\times$ 4

8. 2 1
 $\times$ 4

9. 3 0
 $\times$ 2

10. 3 2
 $\times$ 3

11. 1 1
 $\times$ 3

12. 1 2
 $\times$ 4

13. 3 1
 $\times$ 3

14. 3 2
 $\times$ 2

15. 4 3
 $\times$ 2

16. 2×14

17. 2×22

18. 2×31

19. 3×12

20. 3×21

21. 3×22

25. 2×42

26. 4×22

24. 3×33

PROBLEM SOLVING

25. Manuel ordered furniture for a new office building. He ordered 2 chairs for each of 43 offices. How many chairs did he order? _____

26. A blacksmith made 4 shoes for each of 21 horses. How many horseshoes did he make? _____

Multiplication Models

Name _____

Date _____

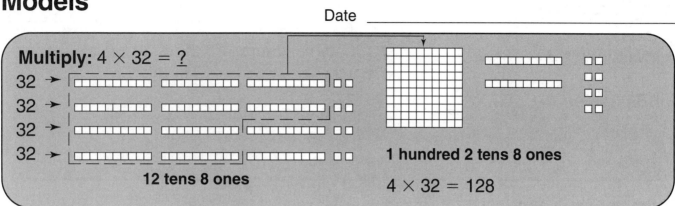

Multiply: 4 × 32 = ?

32 →
32 →
32 →
32 →

12 tens 8 ones

1 hundred 2 tens 8 ones

4 × 32 = 128

Complete. You may use models.

1. 3 × 4 = _____

3 × 40 = _____

2. 6 × 5 = _____

6 × 50 = _____

3. 2 × 9 = _____

2 × 90 = _____

Match the multiplication sentence with the model.

4. 5 × 12 = 60 _____

a.

5. 5 × 13 = 65 _____

b.

6. 3 × 15 = 45 _____

c.

7. 6 × 12 = 72 _____

d.

Use with Lesson 10–3, text pages 326–327.

Multiplying with Regrouping

Name _____

Date _____

Multiply: $6 \times 13 = \underline{?}$	Multiply the ones. Regroup.	Multiply the tens. Then add.
Estimate: $6 \times 10 = 60$	$\begin{array}{r} \overset{1}{13} \\ \times\ 6 \\ \hline 8 \end{array}$	$\begin{array}{r} \overset{1}{1}3 \\ \times\ 6 \\ \hline 78 \end{array}$

Estimate. Then multiply.

1. $\begin{array}{r} 45 \\ \times\ 2 \\ \hline \end{array}$
2. $\begin{array}{r} 49 \\ \times\ 2 \\ \hline \end{array}$
3. $\begin{array}{r} 26 \\ \times\ 3 \\ \hline \end{array}$
4. $\begin{array}{r} 18 \\ \times\ 2 \\ \hline \end{array}$
5. $\begin{array}{r} 15 \\ \times\ 6 \\ \hline \end{array}$

6. $\begin{array}{r} 19 \\ \times\ 4 \\ \hline \end{array}$
7. $\begin{array}{r} 24 \\ \times\ 3 \\ \hline \end{array}$
8. $\begin{array}{r} 48 \\ \times\ 2 \\ \hline \end{array}$
9. $\begin{array}{r} 25 \\ \times\ 3 \\ \hline \end{array}$
10. $\begin{array}{r} 47 \\ \times\ 2 \\ \hline \end{array}$

11. $\begin{array}{r} 24 \\ \times\ 4 \\ \hline \end{array}$
12. $\begin{array}{r} 18 \\ \times\ 5 \\ \hline \end{array}$
13. $\begin{array}{r} 27 \\ \times\ 3 \\ \hline \end{array}$
14. $\begin{array}{r} 13 \\ \times\ 4 \\ \hline \end{array}$
15. $\begin{array}{r} 37 \\ \times\ 2 \\ \hline \end{array}$

16. $\begin{array}{r} \$35 \\ \times\ 2 \\ \hline \end{array}$
17. $\begin{array}{r} \$23 \\ \times\ 4 \\ \hline \end{array}$
18. $\begin{array}{r} \$39 \\ \times\ 2 \\ \hline \end{array}$
19. $\begin{array}{r} 29¢ \\ \times\ 3 \\ \hline \end{array}$
20. $\begin{array}{r} 12¢ \\ \times\ 8 \\ \hline \end{array}$

Find the product.

21. 2×16

22. 2×26

23. 2×38

24. 3×15

25. 3×26

26. 3×28

PROBLEM SOLVING

27. Two buses went to the park. Each bus carried 25 passengers. How many passengers were there in all?

_____

More Multiplying with Regrouping

Name _____

Date _____

Multiply: $4 \times 54 = \underline{?}$	Multiply the ones. Regroup.	Multiply the tens. Then add.
Estimate: $4 \times 50 = 200$	$\begin{array}{r} {}^{1}\\ 5\,4 \\ \times\ 4 \\ \hline 6 \end{array}$	$\begin{array}{r} {}^{1}\\ 5\,4 \\ \times\ 4 \\ \hline 2\,1\,6 \end{array}$

Estimate. Then multiply.

1. $\begin{array}{r} 3\,0 \\ \times\ 4 \\ \hline \end{array}$
2. $\begin{array}{r} 3\,1 \\ \times\ 8 \\ \hline \end{array}$
3. $\begin{array}{r} 6\,0 \\ \times\ 7 \\ \hline \end{array}$
4. $\begin{array}{r} 5\,2 \\ \times\ 3 \\ \hline \end{array}$
5. $\begin{array}{r} 4\,1 \\ \times\ 9 \\ \hline \end{array}$

6. $\begin{array}{r} 3\,8 \\ \times\ 3 \\ \hline \end{array}$
7. $\begin{array}{r} 1\,7 \\ \times\ 6 \\ \hline \end{array}$
8. $\begin{array}{r} 2\,9 \\ \times\ 5 \\ \hline \end{array}$
9. $\begin{array}{r} 1\,6 \\ \times\ 7 \\ \hline \end{array}$
10. $\begin{array}{r} 4\,3 \\ \times\ 4 \\ \hline \end{array}$

11. $\begin{array}{r} 2\,6 \\ \times\ 5 \\ \hline \end{array}$
12. $\begin{array}{r} 6\,3 \\ \times\ 4 \\ \hline \end{array}$
13. $\begin{array}{r} 4\,7 \\ \times\ 3 \\ \hline \end{array}$
14. $\begin{array}{r} 3\,2 \\ \times\ 6 \\ \hline \end{array}$
15. $\begin{array}{r} 5\,3 \\ \times\ 5 \\ \hline \end{array}$

16. $\begin{array}{r} 3\,8 \\ \times\ 4 \\ \hline \end{array}$
17. $\begin{array}{r} 2\,7 \\ \times\ 6 \\ \hline \end{array}$
18. $\begin{array}{r} 8\,3 \\ \times\ 4 \\ \hline \end{array}$
19. $\begin{array}{r} 6\,5 \\ \times\ 4 \\ \hline \end{array}$
20. $\begin{array}{r} 7\,8 \\ \times\ 2 \\ \hline \end{array}$

Write the product.

21. 3×66

22. 8×27

23. 9×34

24. 5×18

25. 6×19

26. 8×32

PROBLEM SOLVING

27. Suzanne fed birds outside her window. She used 4 bags of birdseed each week. How many bags did she use in one year? _____

 119

Multiplying Three Digits

Name_____

Date_____

Multiply: $3 \times 213 = \underline{?}$
Estimate: $3 \times 200 = 600$

Multiply the ones.	Multiply the tens.	Multiply the hundreds.
213	213	213
$\times\quad 3$	$\times\quad 3$	$\times\quad 3$
9	39	639

Estimate. Then multiply.

1. 231
$\times\quad 3$

2. 331
$\times\quad 2$

3. 212
$\times\quad 4$

4. 112
$\times\quad 4$

5. 113
$\times\quad 3$

6. 202
$\times\quad 3$

7. 101
$\times\quad 8$

8. 324
$\times\quad 2$

9. 222
$\times\quad 2$

10. 133
$\times\quad 3$

11. 111
$\times\quad 7$

12. 402
$\times\quad 2$

13. 214
$\times\quad 2$

14. 101
$\times\quad 9$

15. 413
$\times\quad 2$

16. 301
$\times\quad 3$

17. 104
$\times\quad 2$

18. 200
$\times\quad 4$

19. 314
$\times\quad 2$

20. 333
$\times\quad 3$

PROBLEM SOLVING

21. Ms. Diaz ordered supplies for the Lakeview School. She ordered 4 boxes of pencils. Each box held 120 pencils. How many pencils did she order? _____

22. Ms. Diaz ordered craft sticks. She ordered 320 craft sticks for each of 2 classrooms. How many craft sticks did she order altogether? _____

Regrouping in Multiplication

Name_____

Date _____

Multiply: $5 \times 135 = \underline{?}$
Estimate: $5 \times 100 = 500$

Multiply the ones. Regroup.	**Multiply the tens. Add. Regroup again.**	**Multiply the hundreds. Add.**
$\begin{array}{r} {}^{2} \\ 135 \\ \times\ \ 5 \\ \hline 5 \end{array}$	$\begin{array}{r} {}^{1}{}^{2} \\ 135 \\ \times\ \ 5 \\ \hline 75 \end{array}$	$\begin{array}{r} {}^{1}{}^{2} \\ 135 \\ \times\ \ 5 \\ \hline 675 \end{array}$

Write the product.

1. $\begin{array}{r} 238 \\ \times\ \ 3 \\ \hline \end{array}$

2. $\begin{array}{r} 153 \\ \times\ \ 4 \\ \hline \end{array}$

3. $\begin{array}{r} 275 \\ \times\ \ 3 \\ \hline \end{array}$

4. $\begin{array}{r} 167 \\ \times\ \ 5 \\ \hline \end{array}$

5. $\begin{array}{r} 389 \\ \times\ \ 2 \\ \hline \end{array}$

6. $\begin{array}{r} 195 \\ \times\ \ 5 \\ \hline \end{array}$

7. $\begin{array}{r} 169 \\ \times\ \ 4 \\ \hline \end{array}$

8. $\begin{array}{r} 157 \\ \times\ \ 6 \\ \hline \end{array}$

9. $\begin{array}{r} 248 \\ \times\ \ 3 \\ \hline \end{array}$

10. $\begin{array}{r} 134 \\ \times\ \ 7 \\ \hline \end{array}$

11. $\begin{array}{r} 258 \\ \times\ \ 3 \\ \hline \end{array}$

12. $\begin{array}{r} 248 \\ \times\ \ 4 \\ \hline \end{array}$

13. $\begin{array}{r} 146 \\ \times\ \ 6 \\ \hline \end{array}$

14. $\begin{array}{r} 128 \\ \times\ \ 7 \\ \hline \end{array}$

15. $\begin{array}{r} 163 \\ \times\ \ 5 \\ \hline \end{array}$

Estimate. Then multiply.

16. $\begin{array}{r} 270 \\ \times\ \ 3 \\ \hline \end{array}$

17. $\begin{array}{r} 302 \\ \times\ \ 2 \\ \hline \end{array}$

18. $\begin{array}{r} 128 \\ \times\ \ 4 \\ \hline \end{array}$

19. $\begin{array}{r} 193 \\ \times\ \ 3 \\ \hline \end{array}$

20. $\begin{array}{r} 318 \\ \times\ \ 2 \\ \hline \end{array}$

21. $\begin{array}{r} 160 \\ \times\ \ 5 \\ \hline \end{array}$

22. $\begin{array}{r} 163 \\ \times\ \ 4 \\ \hline \end{array}$

23. $\begin{array}{r} 217 \\ \times\ \ 4 \\ \hline \end{array}$

24. $\begin{array}{r} 140 \\ \times\ \ 6 \\ \hline \end{array}$

25. $\begin{array}{r} 136 \\ \times\ \ 7 \\ \hline \end{array}$

PROBLEM SOLVING

26. Marcel goes to a college that is 258 miles from home. When he drives home for a weekend and then returns, how many miles does he drive?

_____

Problem Solving Strategy: Working Backwards

Name_____

Date _____

Alice gave Tim 25 empty bottles for the recycling bin. Then Tim broke 7 bottles. He now has 32 bottles. How many bottles did Tim have to begin with?

To find out how many bottles Tim had at the beginning, start at the end and work backwards.

32	−	25	+	7	=	?
bottles left		bottles from Alice		bottles broke		bottles at the beginning

$32 - 25 + 7 = 14$

Tim had 14 bottles to begin with.

Work backwards to solve. Do your work on a separate sheet of paper.

1. Bart's brother got paid every week. He always kept $85 for groceries and $25 for lunch and busfare. He had $140 left. How much money did he get paid every week?

   _____

2. Yvonne's doctor tells her not to eat for 14 hours before her appointment at 10 A.M. on Tuesday. When is the last time Yvonne can eat before her appointment?

   _____

3. Kyle found 3 baseball cards in his room. His sister gave him 2 more cards. He then had a total of 11 baseball cards. How many cards did he start with?

   _____

4. Monica wants to take the 9:25 train to Philadelphia. The trip from her house to the train station takes 45 minutes. By what time must Monica leave her house?

   _____

5. For Donovan's birthday last week, his sister and brother each gave him 2 games. Then he bought 3 games with his birthday money. Now he has 15 games. How many games did he have to begin with?

   _____

6. Eugene got home at 4:15 P.M. He had walked for 30 minutes from Joe's house, where the two boys spent 1 hour and 15 minutes doing homework. At what time did Eugene and Joe start their homework?

   _____

Division Sense

Name _____

Date _____

Estimate: 29 ÷ 4	Estimate: 79 ÷ 2
$6 \times 4 = 24$ ← too small	2 tens $\times$ 2 = 4 tens ← too small
$7 \times 4 = 28$ ← 29 is between 24 and 32.	3 tens $\times$ 2 = 6 tens ← 7 tens is between 4 tens and 8 tens.
$8 \times 4 = 32$ ← too large	4 tens $\times$ 2 = 8 tens ← too large
Try **7**.	Try **3**. Write zeros for the other digits.
So 29 ÷ 4 is about **7**.	So 79 ÷ 2 is about **30**.

Use facts to estimate.

1. $45 \div 6 =$ _____   **2.** $64 \div 9 =$ _____   **3.** $25 \div 3 =$ _____   **4.** $37 \div 5 =$ _____

5. $52 \div 9 =$ _____   **6.** $48 \div 7 =$ _____   **7.** $15 \div 2 =$ _____   **8.** $26 \div 4 =$ _____

9. $17 \div 2 =$ _____   **10.** $19 \div 3 =$ _____   **11.** $58 \div 8 =$ _____   **12.** $37 \div 7 =$ _____

13. $6\overline{)2\,6}$ **14.** $4\overline{)3\,0}$ **15.** $5\overline{)4\,2}$ **16.** $3\overline{)2\,3}$ **17.** $9\overline{)3\,1}$ **18.** $7\overline{)4\,5}$

Use tens to estimate.

19. $41 \div 2 =$ _____   **20.** $64 \div 5 =$ _____   **21.** $97 \div 4 =$ _____   **22.** $58 \div 3 =$ _____

23. $33 \div 2 =$ _____   **24.** $98 \div 5 =$ _____   **25.** $83 \div 6 =$ _____   **26.** $67 \div 5 =$ _____

27. $92 \div 7 =$ _____   **28.** $84 \div 6 =$ _____   **29.** $75 \div 2 =$ _____   **30.** $88 \div 3 =$ _____

31. $4\overline{)8\,7}$ **32.** $3\overline{)5\,6}$ **33.** $2\overline{)8\,3}$ **34.** $6\overline{)9\,9}$ **35.** $5\overline{)8\,6}$ **36.** $3\overline{)9\,4}$

PROBLEM SOLVING

27. There are 88 stickers. **About** how many stickers can 3 friends share equally? _____

Division with Remainders

Name _____

Date _____

$10 \div 4 = 2$
remainder 2

$10 \div 3 = 3$
remainder 1

$12 \div 5 = 2$
remainder 2

Complete.

1.
$3 \div 2 =$ _____
remainder _____

2.
$8 \div 3 =$ _____
remainder _____

3.
$16 \div 5 =$ _____
remainder _____

4.
$14 \div 5 =$ _____
remainder _____

5.
$11 \div 2 =$ _____
remainder _____

6.
$17 \div 5 =$ _____
remainder _____

7.
$6 \div 4 =$ _____
remainder _____

8.
$15 \div 6 =$ _____
remainder _____

9.
$20 \div 8 =$ _____
remainder _____

Ring circles to show each division. Find the quotient and remainder.

10. $14 \div 6 =$ _____
remainder _____

11. $14 \div 3 =$ _____
remainder _____

12. $19 \div 5 =$ _____
remainder _____

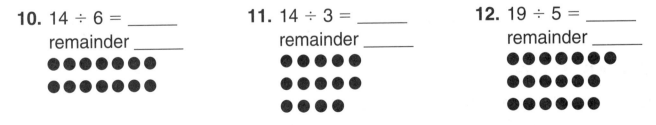

One-Digit Quotients

Name _____

Date _____

Complete. About how many:

1. 5s in 11?   **2.** 5s in 19?   **3.** 2s in 15?   **4.** 3s in 29?   **5.** 4s in 31?

_____   _____   _____   _____   _____

Find the quotient and remainder.

6. $3\overline{)19}$ **7.** $4\overline{)27}$ **8.** $2\overline{)17}$ **9.** $3\overline{)20}$ **10.** $6\overline{)25}$

11. $7\overline{)39}$ **12.** $9\overline{)20}$ **13.** $8\overline{)27}$ **14.** $6\overline{)39}$ **15.** $7\overline{)50}$

Divide and check.

16. $47 \div 6$ **17.** $18 \div 7$ **18.** $57 \div 6$

19. $66 \div 7$ **20.** $37 \div 8$ **21.** $45 \div 8$

PROBLEM SOLVING

22. A snack tray held 13 bags of popcorn. At most,
how many bags of popcorn did each person buy,
if five people bought the same number of bags? _____

Two-Digit Quotients

Name_____

Date _____

Divide.	Multiply.	Subtract and compare.	Bring down the ones. Divide the ones.
$\dfrac{1}{4\overline{)68}}$	$\overset{\times}{\underset{4\overline{)68}}{\overset{1}{}}}\,\overset{1}{\underset{4}{}}$	$\begin{array}{r} 1 \\ 4\overline{)68} \\ -4 \\ \hline 2 \end{array}$	$\begin{array}{r} 17 \\ 4\overline{)68} \\ -4\downarrow \\ \hline 28 \\ -28 \\ \hline 0 \end{array}$

Divide and check.

1. $4\overline{)96}$ **2.** $3\overline{)84}$ **3.** $2\overline{)58}$ **4.** $4\overline{)72}$ **5.** $3\overline{)57}$

6. $5\overline{)95}$ **7.** $4\overline{)52}$ **8.** $3\overline{)81}$ **9.** $2\overline{)98}$ **10.** $5\overline{)75}$

Find the quotient. Show your work on a separate sheet of paper.

11. $82 \div 2 =$ _____   **12.** $55 \div 5 =$ _____   **13.** $92 \div 4 =$ _____

14. $60 \div 6 =$ _____   **15.** $77 \div 7 =$ _____   **16.** $48 \div 4 =$ _____

17. $87 \div 3 =$ _____   **18.** $78 \div 2 =$ _____   **19.** $75 \div 3 =$ _____

PROBLEM SOLVING

20. Hari has 84 glass turtles. If he puts the same number on each of 4 shelves, how many turtles are on each shelf?

_____

21. Katy's 56 banners cover her 4 walls. Each wall has the same number of banners. At most, how many banners are on each wall?

_____

Quotients with Remainders

Name _____

Date _____

$$6\overline{)85} \qquad \overset{\times 1}{6\overline{)85}} \qquad \begin{array}{r} 1 \\ 6\overline{)85} \\ -6 \\ \hline 2 \end{array} \qquad \begin{array}{r} 14 \\ 6\overline{)85} \\ -6\downarrow \\ \hline 25 \\ -24 \\ \hline \end{array} \qquad \begin{array}{r} 14R1 \\ 6\overline{)85} \\ -6 \\ \hline 25 \\ -24 \\ \hline 1 \end{array} \qquad \textbf{Check:} \quad \begin{array}{r} 2 \\ 14 \\ \times 6 \\ \hline 84 \\ +1 \\ \hline 85 \end{array}$$

Divide and check.

1. $3\overline{)52}$ **2.** $4\overline{)49}$ **3.** $4\overline{)67}$ **4.** $5\overline{)97}$ **5.** $6\overline{)82}$

6. $8\overline{)89}$ **7.** $5\overline{)87}$ **8.** $7\overline{)79}$ **9.** $6\overline{)98}$ **10.** $8\overline{)99}$

Find the quotient and remainder.

11. $98 \div 8 =$ _____ **12.** $65 \div 4 =$ _____ **13.** $59 \div 5 =$ _____

14. $89 \div 3 =$ _____ **15.** $49 \div 4 =$ _____ **16.** $74 \div 6 =$ _____

PROBLEM SOLVING

17. Inez has 94 stamps in her collection. She put the same number of stamps into 8 envelopes. At most, how many stamps could be in each envelope? How many would be left over?

_____

Use with Lesson 11–5, text pages 356–357.

Estimating Quotients

Name_____

Date _____

Estimate.

$5.86 ÷ 3 = __?__ $15.37 ÷ 5 = __?__   75 ÷ 4 = __?__

$5.86 ⟶ $6 $15.37 ⟶ $15   75 ⟶ 80

$6 ÷ 3 = $2 $15 ÷ 5 = $3   80 ÷ 4 = 20

So $5.86 ÷ 3 is about $2.   So $15.37 ÷ 5 is about $3.   So 75 ÷ 4 is about 20.

Round to the nearest ten or dollar.

1. 68 _____   **2.** 42 _____   **3.** 85 _____   **4.** 17 _____

5. $1.16 _____   **6.** $3.08 _____   **7.** $24.92 _____   **8.** $17.25 _____

Estimate the quotient.

9. $2\overline{)62}$ **10.** $4\overline{)38}$ **11.** $3\overline{)\$8.90}$ **12.** $6\overline{)\$5.95}$

13. $3\overline{)\$20.99}$ **14.** $6\overline{)\$35.89}$ **15.** $5\overline{)\$14.75}$ **16.** $8\overline{)\$64.25}$

17. $8\overline{)\$63.68}$ **18.** $5\overline{)\$44.79}$ **19.** $6\overline{)\$24.16}$ **20.** $7\overline{)\$35.40}$

4 for $19.75   3 for $27.30   4 for $11.50

2 for $9.55   5 for $9.50

Use the art above to estimate each answer.

21. About how much does 1 tee shirt cost?   _____

22. About how much does 1 belt cost?   _____

23. About how much does 1 pair of shorts cost?   _____

24. How many kites can be bought for $19?   _____

25. How many toy trucks can be bought for $9?   _____

Problem-Solving Strategy: Interpret the Remainder

Name _____

Date _____

There were 19 people camping at the park. Each cabin sleeps 4 people. How many cabins were needed for all 19 people?

Think: What is 19 divided by 4?

$$\overset{4R3}{4\overline{)19}}$$

There were 4 full cabins and 3 people in another cabin. So 5 cabins were needed.

Solve. Do your work on a separate sheet of paper.

1. Lydia's Nursery delivered 65 tree seedlings to the park grounds crew. They planted 9 seedlings at each campsite. How many campsites are there? How many seedlings were left?

   _____

2. Eric spotted 58 geese flying north. They were in 3 groups. Two groups had an equal number of geese and the third group had one more than that. How many geese were in each group?

   _____

3. Isaac wants to clear 50 feet of new trail in 6 days. If he clears about the same number of feet each day, how many feet of trail will he need to clear each day?

   _____

4. There were 37 people who came to the Jones family picnic. Five people came in each car, except for the last car. It held 2 people. How many cars did they use?

   _____

5. A visitor bought 6 park maps. She gave the clerk $50 and received $2 change. How much did each map cost?

   _____

6. A group of 8 campers has 30 marshmallows to share for toasting. Can each camper toast 4 marshmallows?

   _____

7. The road through the park is 23 miles long. Information signs are placed every 5 miles. How many signs are along the road?

   _____

8. Dana has $9. She spent an equal amount on each of 4 postcards and had $1 left. How much did each postcard cost?

   _____

Fractions

Name _____

Date _____

$\frac{3}{4}$ shaded

Numerator ⟶ 3 ⟵ number of equal parts shaded
Denominator ⟶ 4 ⟵ total number of equal parts in the whole or in the set

Word name: three fourths   **Write:** $\frac{3}{4}$

Write the fraction for the shaded part.

1. _____   **2.** _____   **3.** _____   **4.** _____

Write each as a fraction.

5. one fourth _____   **6.** five eighths _____   **7.** three tenths _____

8. four fifths _____   **9.** one third _____   **10.** two sixths _____

11. six twelfths _____   **12.** seven tenths _____   **13.** one half _____

Write the word name for each fraction.

14. $\frac{3}{8}$ _____   **15.** $\frac{9}{10}$ _____

16. $\frac{5}{6}$ _____   **17.** $\frac{2}{5}$ _____

18. $\frac{2}{4}$ _____   **19.** $\frac{7}{8}$ _____

20. $\frac{7}{12}$ _____   **21.** $\frac{4}{9}$ _____

PROBLEM SOLVING

22. Which fraction names the larger part: $\frac{1}{2}$ or $\frac{1}{8}$? _____

Equivalent Fractions

Name_____

Date _____

Equivalent fractions
name the same amount.

$$\frac{2}{3} = \frac{4}{6}$$

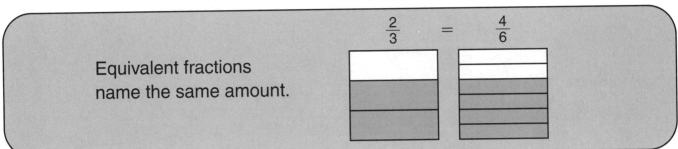

Shade the second figure so that it is equivalent to the first figure. Then complete.

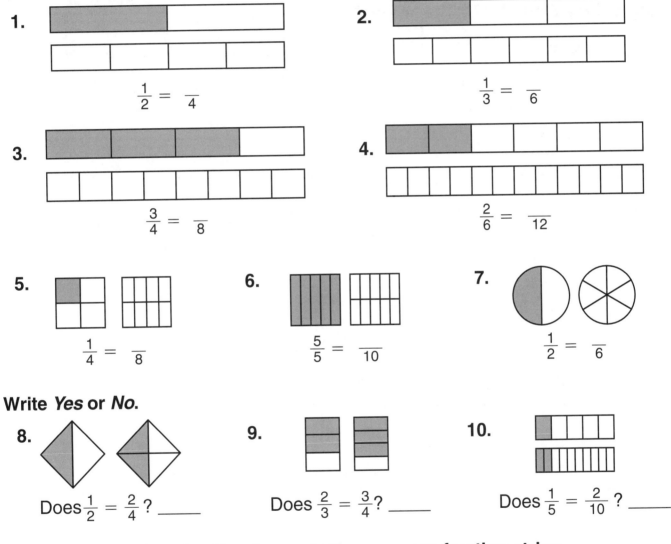

1.

$$\frac{1}{2} = \frac{}{4}$$

2.

$$\frac{1}{3} = \frac{}{6}$$

3.

$$\frac{3}{4} = \frac{}{8}$$

4.

$$\frac{2}{6} = \frac{}{12}$$

5.

$$\frac{1}{4} = \frac{}{8}$$

6.

$$\frac{5}{5} = \frac{}{10}$$

7.

$$\frac{1}{2} = \frac{}{6}$$

Write Yes or No.

8. Does $\frac{1}{2} = \frac{2}{4}$? _____

9. Does $\frac{2}{3} = \frac{3}{4}$? _____

10. Does $\frac{1}{5} = \frac{2}{10}$? _____

Write the equivalent fraction for each. You may use fraction strips.

11. $\frac{1}{3} = \frac{}{9}$

12. $\frac{3}{4} = \frac{}{12}$

13. $\frac{10}{12} = \frac{}{6}$

14. $\frac{6}{10} = \frac{}{5}$

15. $\frac{4}{6} = \frac{}{3}$

16. $\frac{5}{10} = \frac{}{2}$

Estimating Fractions

Name _____

Date _____

Estimating tells you
about how much.

about $\frac{1}{4}$ about $\frac{1}{2}$ about a whole

Is the estimate correct? Write *Yes* or *No*.

1. About what part is full?

Estimate: about $\frac{1}{2}$ _____

2. About how much is written on?

Estimate: about $\frac{1}{4}$ _____

3. About how much is left?

Estimate: about $\frac{1}{4}$ _____

4. About what part of the hour has
passed?

Estimate: about a whole _____

Estimate the fraction for the part of each set that is shaded.
Write *less than* $\frac{1}{2}$ or *more than* $\frac{1}{2}$.

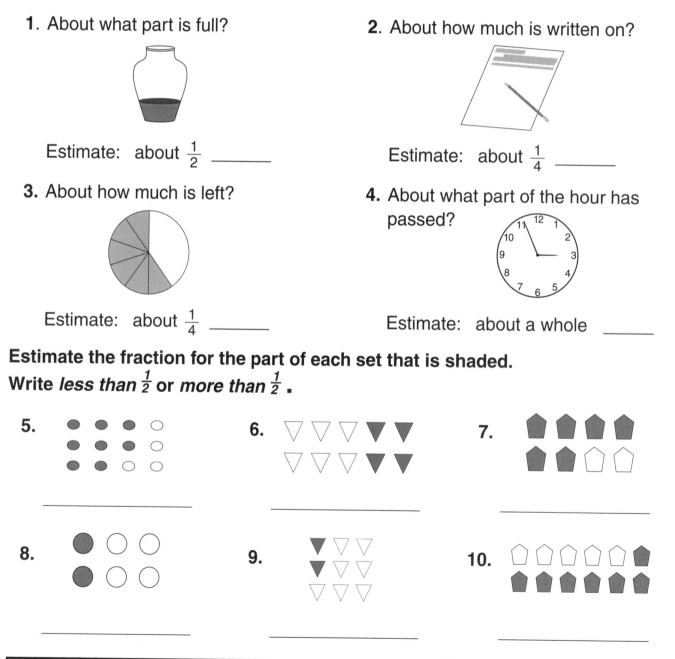

5.
_____

6.
_____

7.
_____

8.
_____

9.
_____

10.
_____

Comparing Fractions

Name _____

Date _____

Compare: $\frac{5}{6}$ ___?___ $\frac{3}{6}$

The denominators are the same.
Compare the numerators. $5 > 3$

$$\frac{5}{6} > \frac{3}{6}$$

Compare $\frac{1}{3}$ ___?___ $\frac{1}{2}$

Look at the models.
Which shows more?

$$\frac{1}{3} < \frac{1}{2}$$

Compare. Write < or >.

1.
   $\frac{3}{8}$ _____ $\frac{5}{8}$

2.
   $\frac{1}{2}$ _____ $\frac{6}{8}$

3.
   $\frac{3}{4}$ _____ $\frac{1}{2}$

4.
   $\frac{1}{4}$ _____ $\frac{1}{2}$

5.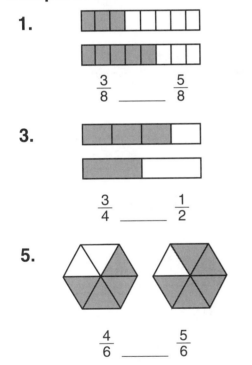
   $\frac{4}{6}$ _____ $\frac{5}{6}$

6.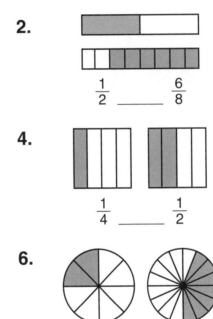
   $\frac{2}{8}$ _____ $\frac{7}{16}$

PROBLEM SOLVING You may use fraction strips.

7. The goal of the third grade was to raise $20 for a new volleyball net. Class A raised $\frac{6}{10}$ of the amount, and Class B raised $\frac{4}{10}$. Which class raised more?

   _____

8. Mary drank $\frac{3}{5}$ of a glass of milk. Peg drank $\frac{2}{5}$ of a glass of milk. Who drank less milk?

   _____

Finding Part of a Number

Name_____

Date_____

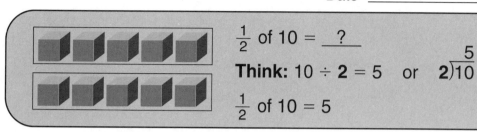

$\frac{1}{2}$ of 10 = __?__

Think: 10 ÷ **2** = 5   or $2\overline{)10}^{\,5}$

$\frac{1}{2}$ of 10 = 5

Write the part of the number.

1. $\frac{1}{2}$ of 4 = _____

2. $\frac{1}{6}$ of 6 = _____

3. $\frac{1}{4}$ of 8 = _____

4. $\frac{1}{3}$ of 9 = _____

Complete.

5. $\frac{1}{2}$ of 6 = _____

6 ÷ 2 = _____

6. $\frac{1}{3}$ of 3 = _____

3 ÷ 3 = _____

7. $\frac{1}{4}$ of 12 = _____

12 ÷ 4 = _____

8. $\frac{1}{2}$ of 12 = _____

12 ÷ 2 = _____

9. $\frac{1}{3}$ of 12 = _____

12 ÷ 3 = _____

10. $\frac{1}{6}$ of 24 = _____

24 ÷ 6 = _____

11. $\frac{1}{2}$ of 8 = _____

12. $\frac{1}{3}$ of 21 = _____

13. $\frac{1}{4}$ of 20 = _____

14. $\frac{1}{10}$ of 50 = _____

15. $\frac{1}{5}$ of 45 = _____

16. $\frac{1}{5}$ of 40 = _____

17. $\frac{1}{8}$ of 24 = _____

18. $\frac{1}{6}$ of 18 = _____

19. $\frac{1}{3}$ of 27 = _____

Mixed Numbers

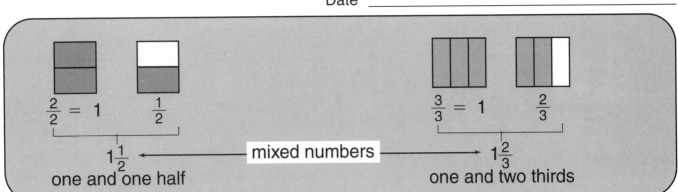

$\frac{2}{2} = 1$ $\frac{1}{2}$ $\frac{3}{3} = 1$ $\frac{2}{3}$

$1\frac{1}{2}$ ⟵ mixed numbers ⟶ $1\frac{2}{3}$

one and one half   one and two thirds

Write the mixed number for each.

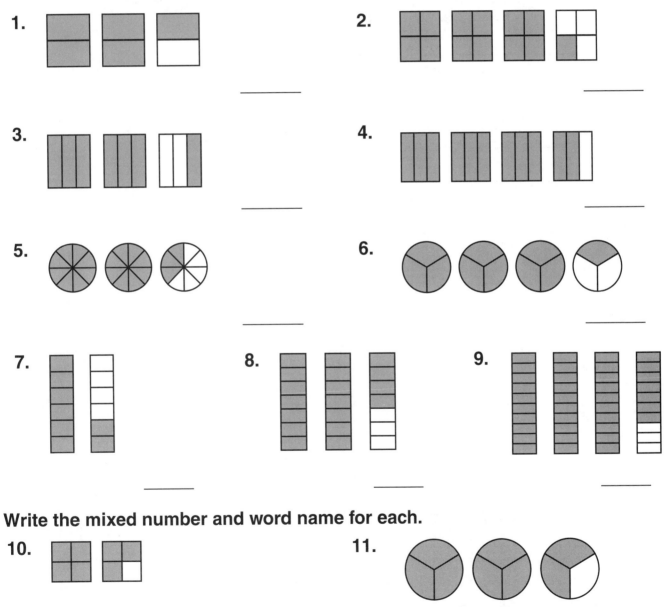

1.

2.

3.

4.

5.

6.

7.

8.

9.

Write the mixed number and word name for each.

10.

11.

    135

Adding Fractions

Name_____

Date _____

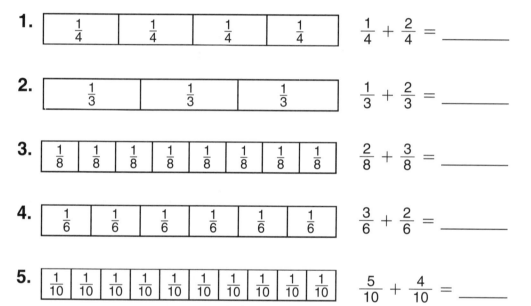

$\frac{2}{5} + \frac{1}{5} = \frac{3}{5}$

Shade each model to find the sum.

1.
| $\frac{1}{4}$ | $\frac{1}{4}$ | $\frac{1}{4}$ | $\frac{1}{4}$ |

$\frac{1}{4} + \frac{2}{4} = $ _____

2.
| $\frac{1}{3}$ | $\frac{1}{3}$ | $\frac{1}{3}$ |

$\frac{1}{3} + \frac{2}{3} = $ _____

3.
| $\frac{1}{8}$ | $\frac{1}{8}$ | $\frac{1}{8}$ | $\frac{1}{8}$ | $\frac{1}{8}$ | $\frac{1}{8}$ | $\frac{1}{8}$ | $\frac{1}{8}$ |

$\frac{2}{8} + \frac{3}{8} = $ _____

4.
| $\frac{1}{6}$ | $\frac{1}{6}$ | $\frac{1}{6}$ | $\frac{1}{6}$ | $\frac{1}{6}$ | $\frac{1}{6}$ |

$\frac{3}{6} + \frac{2}{6} = $ _____

5.
| $\frac{1}{10}$ | $\frac{1}{10}$ | $\frac{1}{10}$ | $\frac{1}{10}$ | $\frac{1}{10}$ | $\frac{1}{10}$ | $\frac{1}{10}$ | $\frac{1}{10}$ | $\frac{1}{10}$ | $\frac{1}{10}$ |

$\frac{5}{10} + \frac{4}{10} = $ _____

Find the sum. You may use fraction strips.

6. $\frac{1}{4} + \frac{1}{4} = $ _____

7. $\frac{1}{10} + \frac{1}{10} = $ _____

8. $\frac{1}{5} + \frac{1}{5} = $ _____

9. $\frac{1}{6} + \frac{4}{6} = $ _____

10. $\frac{3}{8} + \frac{4}{8} = $ _____

11. $\frac{3}{10} + \frac{5}{10} = $ _____

12. $\frac{5}{12} + \frac{2}{12} = $ _____

13. $\frac{1}{8} + \frac{5}{8} = $ _____

14. $\frac{2}{7} + \frac{2}{7} = $ _____

PROBLEM SOLVING You may use fraction strips.

15. Gerene bought $\frac{2}{8}$ yd of blue denim
and $\frac{3}{8}$ yd of red denim. How much
denim did she buy in all?

_____

Subtracting Fractions

Name_____

Date _____

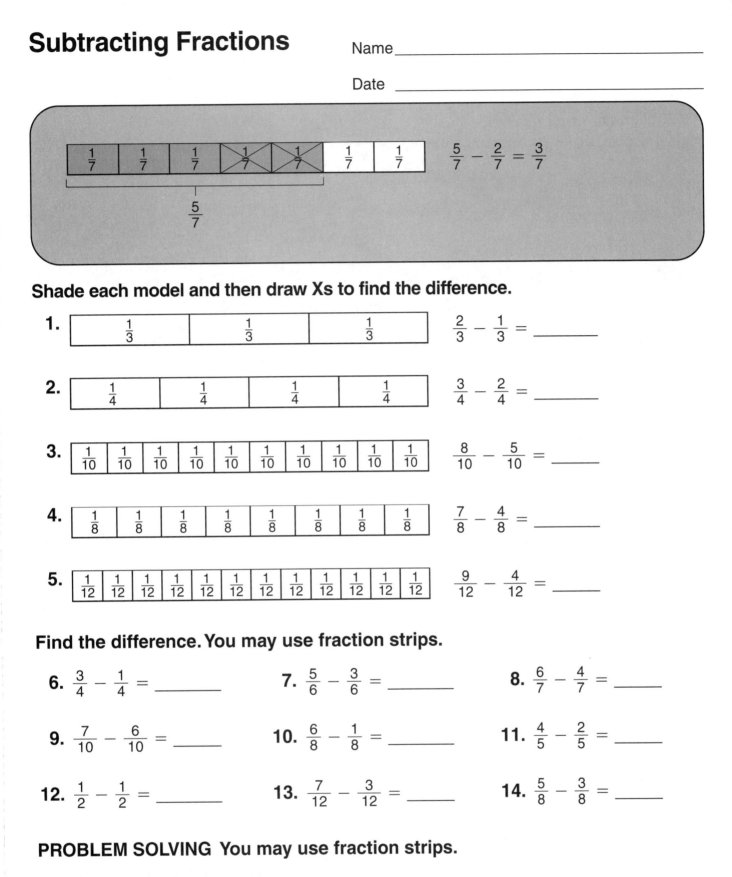

$$\frac{5}{7} - \frac{2}{7} = \frac{3}{7}$$

$$\frac{5}{7}$$

Shade each model and then draw Xs to find the difference.

1.

| $\frac{1}{3}$ | $\frac{1}{3}$ | $\frac{1}{3}$ |

$\frac{2}{3} - \frac{1}{3} =$ _____

2.

| $\frac{1}{4}$ | $\frac{1}{4}$ | $\frac{1}{4}$ | $\frac{1}{4}$ |

$\frac{3}{4} - \frac{2}{4} =$ _____

3.

| $\frac{1}{10}$ | $\frac{1}{10}$ | $\frac{1}{10}$ | $\frac{1}{10}$ | $\frac{1}{10}$ | $\frac{1}{10}$ | $\frac{1}{10}$ | $\frac{1}{10}$ | $\frac{1}{10}$ | $\frac{1}{10}$ |

$\frac{8}{10} - \frac{5}{10} =$ _____

4.

| $\frac{1}{8}$ | $\frac{1}{8}$ | $\frac{1}{8}$ | $\frac{1}{8}$ | $\frac{1}{8}$ | $\frac{1}{8}$ | $\frac{1}{8}$ | $\frac{1}{8}$ |

$\frac{7}{8} - \frac{4}{8} =$ _____

5.

| $\frac{1}{12}$ | $\frac{1}{12}$ | $\frac{1}{12}$ | $\frac{1}{12}$ | $\frac{1}{12}$ | $\frac{1}{12}$ | $\frac{1}{12}$ | $\frac{1}{12}$ | $\frac{1}{12}$ | $\frac{1}{12}$ | $\frac{1}{12}$ | $\frac{1}{12}$ |

$\frac{9}{12} - \frac{4}{12} =$ _____

Find the difference. You may use fraction strips.

6. $\frac{3}{4} - \frac{1}{4} =$ _____

7. $\frac{5}{6} - \frac{3}{6} =$ _____

8. $\frac{6}{7} - \frac{4}{7} =$ _____

9. $\frac{7}{10} - \frac{6}{10} =$ _____

10. $\frac{6}{8} - \frac{1}{8} =$ _____

11. $\frac{4}{5} - \frac{2}{5} =$ _____

12. $\frac{1}{2} - \frac{1}{2} =$ _____

13. $\frac{7}{12} - \frac{3}{12} =$ _____

14. $\frac{5}{8} - \frac{3}{8} =$ _____

PROBLEM SOLVING You may use fraction strips.

15. Hector had $\frac{7}{8}$ cup of milk.
He drank $\frac{4}{8}$ cup of milk.
How much milk was left?

_____

Circle Graphs

Name_____

Date _____

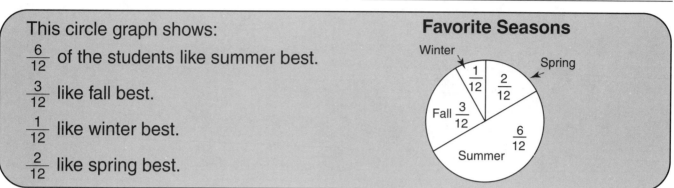

This circle graph shows:

$\frac{6}{12}$ of the students like summer best.

$\frac{3}{12}$ like fall best.

$\frac{1}{12}$ like winter best.

$\frac{2}{12}$ like spring best.

Favorite Seasons

Use the circle graph at the right to answer questions 1–4.

Favorite Activities

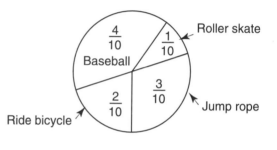

1. What fractional part of the students prefer to jump rope?

   _____

2. What fractional part of the students prefer to roller skate?

   _____

3. What fractional part of the students prefer to play baseball?

   _____

4. Do more students prefer to ride bikes or jump rope?

   _____

Use the circle graph at the right to answer questions 5–8.

Library Selections

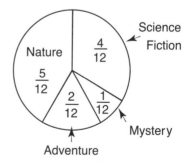

5. What fractional part of the students prefer books about nature?

   _____

6. What fractional part of the students prefer adventure books?

   _____

7. Which type of book has the greatest number of readers?

   _____

8. Which two types of books combined are preferred by half the students?

   _____

Problem-Solving Strategy: Use a Drawing/Model

Name_____

Date_____

Jo drank 6 ounces of a 16-ounce bottle of apple juice. Did she drink more or less than $\frac{1}{2}$ the juice?

Since $\frac{1}{2}$ of 16 = 8, she drank less than $\frac{1}{2}$ of the juice.

Draw a picture:

$\leftarrow \frac{1}{2}$ of 16

6 ounces

PROBLEM SOLVING Do your work on a separate sheet of paper.

1. Peter had 12 baseball cards. He traded $\frac{1}{3}$ of his cards with his cousin and $\frac{1}{4}$ of them with Tyrell. How many cards did he keep?

2. A baker made 32 loaves of bread. One half of the loaves were whole wheat. The rest were rye. How many loaves were rye?

3. Juan's class went on a field trip in a bus. There were 32 seats, but $\frac{1}{4}$ of the seats were empty. How many seats were not empty?

4. Hal's mother is sewing a blouse. She needs 2 yards of fabric. She has a piece of fabric 7 feet long. Does she have enough? (**Hint:** Remember there are 3 feet in 1 yard.)

5. Norm and Liz used two sheets of plywood to make a doll house. They used $\frac{1}{4}$ of a sheet for the floors, $\frac{3}{4}$ of a sheet for the walls, and $\frac{1}{2}$ of a sheet for the roof. How much of the plywood was left?

6. Molly plants 6 rosebushes in $\frac{1}{2}$ of her garden. She plants the same number of tulips, irises, and lilies as rosebushes in the other half of her garden. How many tulips, irises or lilies did she plant?

Fractions and Decimals

Name_____

Date _____

$\frac{1}{10} = 0.1$

one tenth

$\frac{4}{10} = 0.4$

four tenths

$\frac{8}{10} = 0.8$

eight tenths

Write the fraction and the decimal for each.

1.

fraction _____

decimal _____

2.

fraction _____

decimal _____

3.

fraction _____

decimal _____

Write the word name for each.

4. 0.2 _____

5. 0.5 _____

6. 0.9 _____

7. 0.3 _____

8. 0.4 _____

9. 0.7 _____

Write as a decimal.

10. $\frac{3}{10}$ _____

11. $\frac{8}{10}$ _____

12. $\frac{7}{10}$ _____

13. $\frac{6}{10}$ _____

14. $\frac{2}{10}$ _____

15. $\frac{1}{10}$ _____

16. $\frac{5}{10}$ _____

17. $\frac{9}{10}$ _____

PROBLEM SOLVING **Write the answer as a fraction and as a decimal.**

18. The Wolves won seven tenths of their baseball
games. What part of the games did the team win? _____

19. The Lions won five tenths of their games.
What part of the games did the team win? _____

Hundredths

Name_____

Date _____

$$\frac{55}{100} = 0.55 \qquad \text{fifty-five hundredths}$$

Write the fraction and the decimal for each shaded part.

1.

_____

2.

_____

3.

_____

Write as a decimal.

4. $\frac{68}{100}$ _____

5. $\frac{24}{100}$ _____

6 . $\frac{18}{100}$ _____

7. seventy-three hundredths _____

8. fifty hundredths _____

9. eight hundredths _____

10. sixteen hundredths _____

Write as a fraction.

11. 0.71 _____

12. 0.04 _____

13. 0.92 _____

14. 0.88 _____

15. 0.10 _____

16. 0.26 _____

Write the word name for each.

17. $\frac{6}{100}$ _____

18. 0.95 _____

PROBLEM SOLVING **Write the answer as a fraction and as a decimal.**

19. Robin had a box of 100 greeting cards. She
sent 41 cards to her friends during the holidays.
What part of the cards did she send? _____

Decimals Greater Than One

Mixed number: $2\frac{3}{10}$
Decimal: 2.3
Read: two and three tenths

1 whole   1 whole $\frac{3}{10}$

Write the mixed number and the decimal for each.

1.

_____   _____

2.

_____   _____

Write the word name for each.

3. 6.9 _____

4. 7.7 _____

5. 5.4 _____

6. 1.65 _____

7. 9.06 _____

8. 3.12 _____

Write as a decimal.

9. $4\frac{6}{10}$ = _____ **10.** $2\frac{3}{10}$ = _____ **11.** $3\frac{6}{10}$ = _____ **12.** $1\frac{33}{100}$ = _____

13. $8\frac{7}{100}$ = _____ **14.** $1\frac{57}{100}$ = _____ **15.** $6\frac{80}{100}$ = _____ **16.** $3\frac{89}{100}$ = _____

Write *Yes* or *No.*

17. Does $1\frac{8}{10}$ = 1.9? _____ **18.** Does $3\frac{5}{10}$ = 3.5? _____

19. Does $2\frac{2}{10}$ = 2.02? _____ **20.** Does $6\frac{4}{100}$ = 6.04? _____

Comparing and
Ordering Decimals

Name _____

Date _____

0   0.1   0.2   0.3   0.4   0.5   0.6   0.7   0.8   0.9   1.0   1.1   1.2   1.3   1.4   1.5

Compare: 0.4 _?_ 0.7

Think: 0.4 is to the left of 0.7.

0.4 is less than 0.7. So 0.4 $<$ 0.7.

Order from least to greatest: 1.1, 1.0, 1.4

Use the number line: 1.0, 1.1, 1.4

Compare. Write $<$ or $>$. You may draw a number line.

1. 0.3 _____ 0.1

2. 0.2 _____ 0.1

3. 0.7 _____ 0.9

4. $\frac{2}{10}$ _____ 0.6

5. 0.8 _____ $\frac{7}{10}$

6. $\frac{3}{10}$ _____ 0.5

7. $\frac{9}{10}$ _____ 0.1

8. 0.8 _____ $\frac{6}{10}$

9. 2.3 _____ 2.6

10. 1.5 _____ 1.3

11. 3.9 _____ 3.2

12. 7.8 _____ 7.0

13. 6.5 _____ 9.3

14. 8.4 _____ 2.2

15. 3.1 _____ 1.3

Order from least to greatest. You may draw a number line.

16. 0.3, 0.5, 0.1 _____

17. 4.4, 4.5, 4.0 _____

18. 6.1, 6.7, 6.5 _____

PROBLEM SOLVING

19. Marie ran 2.3 miles while training for a race.
David ran 2.7 miles. Who ran the greater distance? _____

20. Brooke walked 0.8 mile on Monday.
She walked 0.6 mile on Wednesday.
On which day did she walk farther? _____

Adding and Subtracting Decimals

Name _____

Date _____

Add:

$$
\begin{array}{r}
\overset{1}{3}.75 \\
+2.53 \\
\hline
6.28
\end{array}
$$

Subtract:

$$
\begin{array}{r}
\overset{5}{\cancel{6}}.\overset{\overset{10}{\cancel{0}}}{\cancel{1}}\overset{15}{\cancel{5}} \\
-3.87 \\
\hline
2.28
\end{array}
$$

Remember to write the decimal point in your answer.

Find the sum.

1. 1.6
    +2.2

2. 7.2
    +1.4

3. 7.4
    +3.7

4. 4.5
    +5.5

5. 8.6
    +4.5

6. 2.5
    +3.7

7. 2 4.7
    +1 3.4

8. 1 4.0 3
    +2 1.9 2

9. 4 1.1 3
    +2 3.3 8

10. 3 4.7 6
    +5 1.6 3

Find the difference.

11. 5.8
    −3.2

12. 8.3
    −2.7

13. 8.6 3
    −4.0 1

14. 8.8 4
    −2.6 3

15. 7.6 2
    −5.3 9

16. 6.7
    −3.4

17. 8.5 0
    −3.4 1

18. 9.3 2
    −1.8 8

19. 5 3.2 7
    −1 4.0 8

20. 6 3.2 6
    −1 2.8 3

Add or subtract. Watch for + and − signs.

21. 2.8 9
    +5.7 0

22. 6.8 0
    −4.3 7

23. 7.1 0
    −2.4 7

24. 9.0 0
    +3.4 8

25. 1.9 0
    +5.5 7

PROBLEM SOLVING

26. Nahoko is training for a cross country race.
She ran 2.8 miles on Saturday and 3.2 miles
on Sunday. How much farther did she run
on Sunday than on Saturday?

_____

Problem-Solving Strategy: Find a Pattern

Name _____

Date _____

What numbers come next in the pattern?

4.2, 4.5, 4.3, 4.6, 4.4, 4.7, 4.5, __?__ __?__

Think: Make a table and look for a pattern.

The next two numbers are 4.8, 4.6.

4.2	4.5	4.3	4.6	4.4
	+ 0.3	− 0.2	+ 0.3	− 0.2

Solve. Do your work on a separate sheet of paper.

1. Yoshi's combination lock must be turned 8 times in a pattern. The first four turns are: 8.1, 7.2, 6.3, 5.4. What are the last four turns?

2. The flight time in a small plane from Cairo to Casablanca is 6.5 h. Every 0.5 h the pilot must report the plane's location. How many times is the location reported?

3. Bettina lifts 3.6 lb the first day, 3.7 lb the second day, 3.9 lb the third day, 4.2 lb the fourth day and 4.6 lb the fifth day, and so on. On what day will she lift *more than* 8 lb?

4. What number comes next in Jack's pattern?

   5.2, 5.0, 5.5, 5.3, 5.8, 5.6, 6.1

5. Bill and Jill sold brownies. The first week they sold 1.5 dozen. Every week they sold 1.5 dozen more than the week before. How many brownies did they sell the fifth week?

6. Sarah can walk one block in 1.2 min. She can run one block in 0.7 min. If she alternates running one block and walking one block, how many blocks can she go in 7.6 min?

Problem Solving:
Review of Strategies

Name_____

Date _____

Solve. Do your work on a separate sheet of paper.

1. Larry, Perry, Jerry, and Mary did a dance in a line. Mary stood before Perry, and Larry was not first in line. In what order did they stand?

_____

2. How many 4-digit numbers between 1000 and 3000 can Marcel write using the digits 0, 1, 2, and 3 only once?

_____

3. Charles built a tree that was 5 feet tall as a prop for the play. Kyle built a tree that was 45 inches tall. How many more inches should Kyle build his tree if both trees need to be the same height?

_____

4. The school play has 2 acts. Each of the two acts lasted for 45 minutes with no intermission. The play ended at 12:30 p.m. What time did the play begin?

_____

5. Alex wanted to buy a book for $5.69 and 3 pens that each cost $.79. He had a $10 bill. Did he have enough money?

_____

6. Daryl cut 6 strips of ribbon for costumes. Each strip needed to be the same length. He had 75 inches of ribbon. How much was left over?

_____

7. During one week Karen delivered 36 newspapers each day. How many newspapers did she deliver in all?

_____

8. Bill wants to set up 115 chairs in equal rows of 8 chairs. How many chairs does he need to fill the last row?

_____

9. Josh divided a number by 8. His quotient was 73. The remainder was 6. What was the number?

_____

10. One factor is 6. The product is 180. What is the missing factor?

_____

Place Value to a Million

Name_____

Date _____

Standard Form: 4,821,350

Read: four million, eight hundred twenty-one thousand, three hundred fifty

Expanded Form: 4,000,000 + 800,000 + 20,000 + 1,000 + 300 + 50

Write the number in standard form.

1. 3 million _____

2. six million _____

3. 5,000,000 + 700,000 _____

4. 2,000,000 + 100,000 + 400 _____

5. 6,000,000 + 900,000 + 40,000 + 3000 + 200 + 60 _____

6. four million, two hundred sixty-three thousand, five hundred _____

7. one million, three hundred twenty thousand, nine hundred fifty _____

Complete.

8. 5,000,000 = _____ millions

9. 2,000,000 = _____ millions

10. 3,490,127 = _____ millions _____ hundred thousands

_____ ten thousands _____ thousands _____ hundred

_____ tens _____ ones

11. 5,060,638 = _____ millions _____ hundred thousands

_____ ten thousands _____ thousands _____ hundreds

_____ tens _____ ones

12. 4,197,052 = _____ millions _____ hundred thousand

_____ ten thousands _____ thousands _____ hundreds

_____ tens _____ ones

In what place is the underlined digit? What is its value?

13. 8,8<u>0</u>9,151 _____

14. <u>2</u>,043,600 _____

_____

_____

Add and Subtract Larger Numbers

Name_____

Date _____

Add: 37,531 + 21,873

$$\begin{array}{r} \overset{1}{3}\,7,\overset{1}{5}\,3\,1 \\ +2\,1,8\,7\,3 \\ \hline 5\,9,4\,0\,4 \end{array}$$

Subtract: 52,534 − 27,680

$$\begin{array}{r} 5\,2,5\,3\,4 \\ -2\,7,6\,8\,0 \\ \hline 2\,4,8\,5\,4 \end{array}$$

Add or subtract. Watch for + and − .

1. 43,621
    +21,760

2. 37,569
    +43,892

3. 72,437
    −13,119

4. 95,300
    −43,296

5. 45,052
    −21,782

6. 37,118
    +56,938

7. 64,502
    +20,924

8. 23,866
    − 1,572

9. $362.18
    + 114.06

10. $230.00
    − 68.90

11. $767.22
    − 495.25

12. $894.63
    + 57.92

Align and add or subtract.

13. 37,625 − 18,483

14. 24,095 + 30,547

15. $385.95 − $47.60

16. $492.99 + $65.00

PROBLEM SOLVING

17. The average depth of the Pacific Ocean is 12,925 feet.
The average depth of the Atlantic Ocean is 11,730 feet.
How much deeper is the Pacific Ocean? _____

Divisibility

Name_____

Date _____

$10 \div 2 = 5$	$10 \div 5 = 2$	$10 \div 10 = 1$
Any number ending in 0, 2, 4, 6, or 8 is *divisible* by 2.	Any number ending in 0 or 5 is *divisible* by 5.	Any number ending in 0 is *divisible* by 10.

Is the number divisible by 2? Write *Yes* or *No*.

1. 4 _____   **2.** 35 _____   **3.** 40 _____   **4.** 59 _____

5. 213 _____   **6.** 754 _____   **7.** 1002 _____   **8.** 6365 _____

Is the number divisible by 5? Write *Yes* or *No*.

9. 20 _____   **10.** 32 _____   **11.** 55 _____   **12.** 67 _____

13. 395 _____   **14.** 521 _____   **15.** 1080 _____   **16.** 4489 _____

Is the number divisible by 10? Write *Yes* or *No*.

17. 15 _____   **18.** 50 _____   **19.** 48 _____   **20.** 99 _____

21. 205 _____   **22.** 330 _____   **23.** 7700 _____   **24.** 1001 _____

Complete the table. Write *Yes* or *No*.

	Divisible by	20	35	68	92	110	3152
25.	2						
26.	5						
27.	10						

PROBLEM SOLVING

28. I am a number that is divisible by 2 and 10. I am between 35 and 45. What number am I?   _____

Missing Digits

Name_____

Date _____

Use guess and test.

$1 + 7 + ? = 9$ →
$$
\begin{array}{r}
7\ \square \\
+\ \square\ 7 \\
\hline
9\ 6 \\
\end{array}
$$
← $? + 7 = 16$

$$
\begin{array}{r}
^{1}\ \ \\
7\ 9 \\
+1\ 7 \\
\hline
9\ 6 \\
\end{array}
$$

$? - 2 = 0$

$$
\begin{array}{r}
3\ \square\ 8 \\
\end{array}
$$

$3 - ? = 2$ →
$$
\begin{array}{r}
3\ \square\ 8 \\
-\ \square\ 2\ \square \\
\hline
2\ 0\ 1 \\
\end{array}
$$
→ $8 - ? = 1$

$$
\begin{array}{r}
3\ 2\ 8 \\
-1\ 2\ 7 \\
\hline
2\ 0\ 1 \\
\end{array}
$$

Write the missing digits.

1.
$$
\begin{array}{r}
2\ \underline{} \\
+\ \underline{}\ 8 \\
\hline
4\ 9 \\
\end{array}
$$

2.
$$
\begin{array}{r}
4\ \underline{} \\
+\ \underline{}\ 5 \\
\hline
6\ 8 \\
\end{array}
$$

3.
$$
\begin{array}{r}
8\ \underline{} \\
-\ \underline{}\ 6 \\
\hline
3\ 1 \\
\end{array}
$$

4.
$$
\begin{array}{r}
\underline{}\ 2 \\
-\ 6\ \underline{} \\
\hline
3\ 0 \\
\end{array}
$$

5.
$$
\begin{array}{r}
3\ \underline{}\ 0 \\
+\ \underline{}\ 5\ \underline{} \\
\hline
7\ 7\ 3 \\
\end{array}
$$

6.
$$
\begin{array}{r}
\underline{}\ \underline{}\ 2 \\
+\ 3\ 4\ \underline{} \\
\hline
8\ 8\ 3 \\
\end{array}
$$

7.
$$
\begin{array}{r}
6\ \underline{}\ 6 \\
-\ \underline{}\ 4\ \underline{} \\
\hline
4\ 1\ 3 \\
\end{array}
$$

8.
$$
\begin{array}{r}
6\ 9\ \underline{} \\
-\ \underline{}\ 3\ 3 \\
\hline
4\ 6\ 2 \\
\end{array}
$$

9.
$$
\begin{array}{r}
4\ \underline{} \\
\times\ \ \ 4 \\
\hline
1\ 6\ 4 \\
\end{array}
$$

10.
$$
\begin{array}{r}
\underline{}\ 3 \\
+\ 4\ \underline{} \\
\hline
1\ 1\ 0 \\
\end{array}
$$

11.
$$
\begin{array}{r}
\underline{}\ 1 \\
-\ 3\ \underline{} \\
\hline
1\ 2 \\
\end{array}
$$

12.
$$
\begin{array}{r}
4\ \underline{} \\
\times\ \ \ 8 \\
\hline
3\ 7\ 6 \\
\end{array}
$$

13.
$$
\begin{array}{r}
\underline{}\ 2\ 0\ \underline{} \\
+\ 1\ \underline{}\ \underline{}\ 3 \\
\hline
6\ 8\ 9\ 8 \\
\end{array}
$$

14.
$$
\begin{array}{r}
6\ \underline{}\ 0\ \underline{} \\
-\ \underline{}\ 1\ \underline{}\ 3 \\
\hline
4\ 3\ 7\ 6 \\
\end{array}
$$

15.
$$
\begin{array}{r}
2\ \underline{}\ \underline{}\ 4 \\
-\ \ \ 3\ 6\ 8 \\
\hline
\underline{}\ 9\ 2\ 6 \\
\end{array}
$$

PROBLEM SOLVING

16. Lori played a computer game with a missing digit. Help her find the missing digit. _____

$$
\begin{array}{r}
4\ \underline{} \\
\times\ \ \ 4 \\
\hline
1\ 8\ 0 \\
\end{array}
$$

---

150 **Use with Lesson 14–4, text pages 424–425.**

Copyright © William H. Sadlier, Inc. All rights reserved.

Order of Operations

Solve 3 + 2 × 4:

Multiply or divide in order from left to right.
Add or subtract in order from left to right.

So 3 + 2 × 4 = 11.

$$3 + 2 \times 4$$
$$3 + \quad 8$$
$$11$$

Write the letter of the operation that should be done first. Then compute.

a. addition   **b.** subtraction   **c.** multiplication   **d.** division

1. 7 + 4 − 2   _____   **2.** 63 − 2 + 5   _____

3. 53 − 4 × 2   _____   **4.** 15 ÷ 5 × 2   _____

5. 3 + 4 − 1 × 6   _____   **6.** 8 − 2 × 3 ÷ 3   _____

7. 5 × 3 + 7 =   _____   **8.** 6 + 8 × 2 =   _____

9. 6 + 6 ÷ 2 =   _____   **10.** 12 ÷ 3 + 4 =   _____

11. 36 − 15 ÷ 5 =   _____   **12.** 48 − 36 ÷ 4 =   _____

13. 10 ÷ 2 + 9 ÷ 1 =   _____   **14.** 32 ÷ 4 − 4 ÷ 2 =   _____

PROBLEM SOLVING

15. Use each operational symbol once
to solve each sentence.

12 ◯ 4 ◯ 2 = 5

12 ◯ 4 ◯ 2 = 4

Use with Lesson 14–5, text pages 426–427.

151

Missing Operation

Name _____

Date _____

Use guess and test to find the operation.

$6 \underline{\;?\;} 9 = 54$	$12 \underline{\;?\;} 7 = 5$
Think: 54 is greater than both 6 and 9.	**Think:** 5 is less than both 7 and 12.
Test addition and multiplication.	Test subtraction and division.
Test: $6 + 9 = 54$ not true	**Test:** $12 \div 7 = 5$ not true
$\;\; 6 \times 9 = 54$ true	$\;\; 12 - 7 = 5$ true

Write + or − to complete.

1. $8 \underline{\quad} 5 = 13$ **2.** $6 \underline{\quad} 6 = 12$ **3.** $13 \underline{\quad} 6 = 7$ **4.** $7 \underline{\quad} 7 = 0$

5. $6 \underline{\quad} 4 = 10$ **6.** $6 \underline{\quad} 3 = 9$ **7.** $5 \underline{\quad} 5 = 10$ **8.** $8 \underline{\quad} 2 = 10$

9. $9 \underline{\quad} 1 = 10$ **10.** $14 \underline{\quad} 8 = 6$ **11.** $9 \underline{\quad} 6 = 15$ **12.** $17 \underline{\quad} 9 = 8$

Write × or ÷ to complete.

13. $5 \underline{\quad} 8 = 40$ **14.** $56 \underline{\quad} 7 = 8$ **15.** $6 \underline{\quad} 4 = 24$ **16.** $2 \underline{\quad} 9 = 18$

17. $3 \underline{\quad} 3 = 1$ **18.** $35 \underline{\quad} 7 = 5$ **19.** $4 \underline{\quad} 7 = 28$ **20.** $1 \underline{\quad} 7 = 7$

21. $30 \underline{\quad} 6 = 5$ **22.** $18 \underline{\quad} 3 = 6$ **23.** $9 \underline{\quad} 5 = 45$ **24.** $0 \underline{\quad} 8 = 0$

Write +, −, × or ÷ to complete.

25. $14 \underline{\quad} 7 = 7$ **26.** $8 \underline{\quad} 4 = 12$ **27.** $4 \underline{\quad} 4 = 16$ **28.** $12 \underline{\quad} 3 = 4$

29. $9 \underline{\quad} 8 = 1$ **30.** $2 \underline{\quad} 4 = 8$ **31.** $42 \underline{\quad} 7 = 6$ **32.** $8 \underline{\quad} 8 = 64$

PROBLEM SOLVING

33. Margaret's swim team practices for 2 hours each day from Monday to Friday. How many hours do they practice in one week?

_____

Factors

Name_____

Date _____

> Use multiplication sentences to find the factors
> of a number and the common factors of
> two or more numbers.
>
> | Factors of 8: **1, 2**, 4, 8 | Factors of 6: **1, 2**, 3, 6 | Common factors |
> |---|---|---|
> | $1 \times 8 = 8$ | $1 \times 6 = 6$ | of 6 and 8: |
> | $2 \times 4 = 8$ | $2 \times 3 = 6$ | **1, 2** |

Write the missing factor.

1. ___ $\times\ 10 = 10$ **2.** ___ $\times\ 5 = 10$ **3.** ___ $\times\ 12 = 12$ **4.** ___ $\times\ 6 = 12$

5. ___ $\times\ 4 = 12$ **6.** ___ $\times\ 6 = 12$ **7.** ___ $\times\ 4 = 4$ **8.** ___ $\times\ 2 = 4$

List all the factors of each.
You may use multiplication sentences.

9. 24 _____

10. 18 _____

11. 20 _____

12. 32 _____

13. 36 _____

List all the common factors of each set of numbers.

14. 12 and 16 _____   **15.** 8 and 12 _____

16. 20 and 40 _____   **17.** 15 and 30 _____

18. 9 and 24 _____   **19.** 10 and 40 _____

20. 16 and 24 _____   **21.** 18 and 30 _____

   153

Multiplying Money

Name _____

Date _____

What is the total cost of 4 pairs of socks?

Estimate: $4 \times \$2.00 = \$8.00$

$2.39

Multiply:

$$
\begin{array}{r}
{}^{1\ 3}\\
\$2.39\\
\times\quad 4\\
\hline
\$9.56
\end{array}
$$

Write $ and .
in the product.

Four pairs of socks cost $9.56.

Estimate. Then multiply.

1. $.72
    × 4

2. $.37
    × 2

3. $2.83
    × 3

4. $3.60
    × 2

5. $1.74
    × 4

6. $.45
    × 5

7. $.28
    × 6

8. $4.53
    × 3

9. $2.76
    × 6

10. $5.69
    × 4

Write the product.

11. $.60
    × 3

12. $.75
    × 4

13. $1.85
    × 5

14. $3.56
    × 6

15. $4.89
    × 3

16. $.78
    × 2

17. $.98
    × 4

18. $3.80
    × 5

19. $2.54
    × 6

20. $1.97
    × 7

PROBLEM SOLVING

21. Maidee sold 6 large shell bracelets
    for $2.45 each. What was the total cost? _____

22. Jason sold 8 small bracelets
    for $1.98 each. What was the total cost? _____

Dividing Money

Name _____

Date _____

How much does one apple cost?

$1.80 for 6

Divide:

$$\begin{array}{r} \$\ .30 \\ 6\,\overline{)\$1.80} \\ \underline{-1\,8} \\ 0 \\ \underline{-0} \\ 0 \end{array}$$

Write $ and . in the quotient above $ and . in the dividend.

One apple costs $.30.

Complete.

1. $\dfrac{\$.4}{2\,\overline{)\$.8\,8}}$

2. $\dfrac{\$.2}{3\,\overline{)\$.6\,9}}$

3. $\dfrac{\$.3}{2\,\overline{)\$.6\,0}}$

4. $\dfrac{\$\ .2}{5\,\overline{)\$1.0\,0}}$

5. $\dfrac{\$\ \ }{2\,\overline{)\$2.4\,0}}$

6. $\dfrac{\$\ \ }{3\,\overline{)\$9.3\,9}}$

7. $\dfrac{\$\ \ }{4\,\overline{)\$2.4\,4}}$

8. $\dfrac{\$\ \ }{5\,\overline{)\$1\,0.0\,0}}$

Divide and check.

9. $2\,\overline{)\$.4\,8}$

10. $5\,\overline{)\$.6\,0}$

11. $4\,\overline{)\$9.1\,6}$

12. $5\,\overline{)\$2.7\,5}$

13. $6\,\overline{)\$8.4\,0}$

14. $3\,\overline{)\$8.3\,4}$

15. $4\,\overline{)\$4.9\,6}$

16. $6\,\overline{)\$9.7\,2}$

17. $\$.96 \div 8 = $ _____

18. $\$.70 \div 5 = $ _____

19. $\$4.00 \div 2 = $ _____

20. $\$6.30 \div 3 = $ _____

21. $\$9.12 \div 6 = $ _____

22. $\$8.56 \div 4 = $ _____

PROBLEM SOLVING

23. Three T-shirts come in a package that costs $6.75. What is the cost of one T-shirt?

_____

24. Eight headbands come in a bag that costs $7.60. What is the cost of one headband?

_____

Problem-Solving Strategy: More Multi-Step Problems

Name _____

Date _____

Arlene wanted to make fruit salad for the party. She bought apples for $7.18, bananas for $6.42 and pears for $9.77. How much change did she get from $30.00?

Arlene's change was $6.63.

Think: Add $7.18 + $6.42 + $9.77. Subtract the sum from $30.00 to find how much Arlene has left.

```
  1  1                9  9
$7.1 8            2 10 10 10
  6.4 2           $3 0. 0 0
+ 9.7 7           - 2 3. 3 7
$2 3. 3 7           $  6. 6 3
```

Solve. Do your work on a separate sheet of paper.

1. Daisy sold 9 apple dumplings for a total of $11.25. Each dumpling cost $.75 to make. How much profit did Daisy make from selling her dumplings?

_____

2. Melissa scored 48 points in a video game. Then she doubled her points. Andre scored 31 points and then tripled his points. Who had a higher score?

_____

3. Cherries are $1.29 a pound. Randy buys 5 pounds of cherries to make jelly. How much change will he get from a ten-dollar bill?

_____

4. A rectangular sign is twice as long as it is wide. It is 15 feet wide. What is the perimeter of the sign?

_____

5. Cereal X has 3 times more oats than Cereal Y, which has 2 times more oats than Cereal Z. If Cereal X has 30 ounces of oats, how many ounces of oats are in Cereal Z?

_____

6. Miguel needs to earn $35.00 for a camping trip. If he mows lawns for 3 days and earns $7.95 each day and rakes leaves for 2 days at $5.00 per day, will he earn enough money?

_____